HANDBOOK OF MEASURING UNITS AND CONVERSIONS

HANDBOOK OF MEASURING UNITS AND CONVERSIONS

DR. R. P. KACHRU
Assistant Director General
(Process Engineering)
Indian Council of Agricultural Research
New Delhi - 110 001

DR. G. SINGH
Director
Central Institute of Agricultural Engineering
Bhopal

DR. A. K. JAIN
Senior Scientist
Indian Council of Agricultural Research
New Delhi - 110 001

ANMOL PUBLICATIONS PVT. LTD.
NEW DELHI-110002 (INDIA)

ANMOL PUBLICATIONS PVT. LTD.
H.O.: 4374/4B, Ansari Road, Darya Ganj
New Delhi-110 002 (India)
Ph.: 23278000, 23261597
B.O.: 1015, Ist Main Road, BSK IIIrd Stage
IIIrd Phase, IIIrd Block
Bangalore - 560 085 (Karnataka)
Ph: 080-41723429, Tel/Fax: 080-2672 3604
Visit us at: www.anmolpublications.com

Handbook of Measuring Units and Conversions

First Published 1999

Reprint 2007

PRINTED IN INDIA

Printed at Mehra Offset Press, Delhi.

Contents

Foreword

Different systems of units of measurements are adopted world over for scientific activities and trade transactions. The length, area, volume and weight were the major quantities measured in the early days of civilization. With industrialization and scientific advancement, particularly, when developments like space travel has become reality, measurements of other quantities such as for light, electricity, magnetism, radiation etc. became important. The FPS, CGS and MKS systems of units of measurements, at one time, were quite popular amongst scientists and engineers for quantification of values till Systéme International d'Unites (SI) was universally accepted for adoption in 1960. As a result, the conversion of values from one system of units of measurement to another system became complex. The Bureau of Standards in different countries look after the development and adoption of units of measurements. Since international cooperation in science and technology has become important vis-a-vis the globalization of trade and explosion of information technology, the understanding of measuring units adopted in different countries has become vital.

The publication of Hand Book of Measuring Units and Coversions by Drs. R.P. Kachru, G Singh and A. K. Jain is in the direction of facilitating the understanding of units of measurement and conversions of values from one system of units to another. Incorporating all the systems of units of measurements adopted world over is not a feasible proposition, but the authors have tried their best to incorporate most of the units of measurements. The publication is expected to have universal acceptance by the scientists, engineers and trade promoters. The authors deserve the appreciation in bringing out such a comprehensive Hand Book of Measuring Units and Conversions, and I congratulate them for this effort. I hope the readers will find this document very useful in understanding the units of measurements adopted world over and their conversions to other systems.

(R.S. PARODA)
Secretary,
Department of Agricultural Research and Education (DARE)
Government of India and
Director General,
New Delhi Indian Council of Agricultural Research (ICAR)

Preface

Countries, all over the world, had practiced various methods of barter trade system which lead to development of units of measurements. The units of measurements adopted by different countries varies not only from country to country, but also from region to region within the country, like chain, gudge, perch, mou, bigha, oke, bushel, pint, pound, quintal, seer, varas etc. Over the period of time, however, FPS, CGS and MKS system of units of measurements became very common, especially with the scientists and engineers. The adoption of Metric System or Systéme International d'Unites (SI) of measurements internationally by the scientists and engineers led to the initial problems of understanding and comparing the values with local system of units of measurements. Even scientists and engineers following FPS, CGS and MKS systems found it difficult to convert the actual quantities to SI units. The SI, however, brought in standardization of units of measurements.

The hand book has been compiled to facilitate its usage as a ready reference material for conversion of units from one system to another. The publication has been divided in five chapters, covering topics on units & measurements—historical developments; SI units; units, dimensions & conversion methods; some selected units of measurements which are/were used in different countries and their equivalents and the last chapter on conversion factors. Efforts have been made to enlist all the units of measurements in various systems adopted in different countries, but there may be some inadvertent omissions.

Any additional information and suggestions from the readers are welcome and would be very much helpful in future updation of the book. The authors are greatly indebted to Dr. R.S. Paroda,

Secretary, Department of Agricultural Research and Education (DARE) and Director General, Indian Council of Agricultural Research (ICAR) for going through the manuscript and writing "Foreword" for it. We believe that the users will find the Hand Book extremely helpful in their scientific references as well as in trade transactions.

New Delhi
November 1998

R.P. Kachru
G. Singh
A.K. Jain

1

Units & Measurements—Historical Developments

1.1 Introduction

The concept of weights and measures in the early days of civilization originated from the need for some kind of uniformity in exchanging things (bartering) leading to development of primitive trade and commercial activities. Initially, the concept was confined to mass or weight (a measure for amount of solids), volume (a measure for liquids), area (a measure for land coverage) and length (a measure for distances). Later, the concept was extended to time, temperature, luminosity, electric current and molecular mass.

The essence of the concept of weights and measures are the concept of (1) Uniformity, (2) Units and (3) Standards. *Uniformity*, a necessity for any system of weights and measures, requires accuracy, reliability and repeatability. *Unit* is the name given to a fixed quantity which can be repeated. *Standard* is the physical embodiment of a Unit kept for its accurate and reliable repetition.

Evolution of civilization has led to haphazard growth of system of weights and measures based on local customs and need of the hour. A planned system of units came into being only in October 1960 during the 11th General Conference on Weights and Measures, in Paris. This modern scientific system of units and measurement is known as 'Systéme International d' Unités' or SI units.

1.1.1 Unit

A Unit is the yardstick by which we measure the sizes of a physical characteristic of a system. It may be thought as the standard of comparison, e.g., m, g, sec etc. which is arbitrarily established. Both the unit and the measuring procedure must be defined by some pre-specified operational procedure. Such a procedure must have three things :

1. a datum
2. a unit of measure defined by an operational procedure
3. an operational rule for interpolation and extrapolation

In the case of length, the datum is taken to be zero; the unit of measure is the metre as established by comparison with a standard metre; and the rule of interpolation is given by means of fractions which are marked on the measuring device.

To specify and to carry out calculations with physical quantities, we have to define them in *kind* and in *magnitude*. We do this by expressing each quantity as a multiple, or fraction, of some yardstick or other standard measure of the same kind of quantity. We call the standard measure of each kind of quantity the *unit* and the number of times the unit is contained in any given amount of the same kind of quantity the *number of measure*. The kind of physical quantity, as well as the size of the yardstick used for its measurement is defined by the *Unit*. Without unit, the number of measure has no physical meaning.

If Q is used to denote some physical quantity, then, when thought of in terms of some unit

$$Q = \{Q\} [Q]$$

where, [Q] denotes *Unit* and {Q} *number of measure* of Q in terms of that unit.

The magnitude of any given physical quantity is, of course, independent of the size of the unit employed for its measurement, the choice of the latter affecting merely the number of measure accordingly.

In the general case, if any given quantity Q, is measured first in terms of some *unit* [Q], and again in terms of *different size of unit* [Q]', then

$$Q = \{Q\} [Q] = \{Q\}' [Q]'$$

$$\therefore \quad \{Q\}'/\{Q\} = [Q]/[Q]'$$

$$\text{or} \quad \{Q\}'=\{Q\}[Q]/[Q]' = \text{Constant}/[Q]'$$

The above equation tells that the number of measure is always inversely proportional to the size of the unit.

Further, if any two given physical quantities of the same kind Q_1 and Q_2, are measured first in terms of some unit [Q], and, again in terms of a different size of unit [Q]', then,

$$Q_1 = \{Q_1\} [Q] \quad = \{Q_1\}' [Q]' \tag{1}$$

$$Q_2 = \{Q_2\} [Q] \quad = \{Q_2\}' [Q]' \tag{2}$$

Dividing the (1) by (2),

$$Q_1/Q_2 \quad = \{Q_1\}/\{Q_2\} = \{Q_1\}'/\{Q_2\}'$$

It is seen that the units have vanished, leaving only the ratios of the numbers of measures, and moreover, that the two ratios of the numbers of measure are equal. In other words, the ratio Q_1/Q_2 of any two given physical quantities of the same kind, is a fixed number, independent of the size of the units employed for the measurement of Q_1 and Q_2, that number being the number of measure of Q_1 in terms of Q_2.

1.1.2 Measurements

Measurement presupposes something to be measured, and, unless we know what that something is, no measurement can have any significance.

Definition and measurement certainly have functional similarities which make it almost inevitable that a discussion of one should sooner or later involve the other. They both have the character of leading to relations which set the entities of science in order with respect to one another. The kinds of order that they establish can be broadly differentiated, but they run together in many cases, so that there are times when the two procedures seem to amount almost to the same thing. Definition, in general, is concerned with the systematic order of the conceptual schemes of science, and with the nature of the relations between different entities.

Measurement has a more limited function; that of establishing metrical order among different manifestations of particular properties, and of making scientific events amenable to mathematical description. "*Measurement* is the assignment of particular mathematical characteristics to conceptual entities in such a way as to permit (1) an unambiguous mathematical description of every situation involving the entity and (2) the arrangement of all occurrences of it in a quasi-serial order." "Quasi-serial order" in Hempels' sense is to mean an order which determines, for any two occurrences, either that they are equivalent with respect to the property in question that one is greater than the other.

Under the modern view, the process of measurement is the process of mapping empirical properties or relations into a formal model. Measurement is possible only because there is a kind of isomorphism between (1) the empirical relations among properties of objects and events and (2) the properties of the formal game in which numerals are the pawns and operators the moves.

As Russell put it "*Measurement* demands some one-one relation between the numbers and magnitudes in question - a relation which may be direct or indirect, important or trivial, according to circumstances". The classical view of measurement, as Campbell represents it, is essentially the view that direct or "fundamental" measurement is possible only when the "axioms of additivity" can be shown to be isomorphic with the manipulations we perform upon objects. Only a few properties, such as length, weight, and electric resistance, are measurable in this fundamental way. Most other magnitudes are measured by indirect or "derived" measurement—process in which derived magnitudes are defined by means of numerical laws relating fundamental magnitudes (e.g.; density etc.)

To make it clear that the classical terms of "*fundamental*" and "*derived*" describe two classes of operations, not two classes of scales. Both fundamental measurement and derived measurement usually result in ratio scales, which are invariant under multiplication by a constant, or, as is sometimes said, are unique up to a similarity transformation.

1.2 Development of Systems of Units and Measurements

Early weight and volume measurements may have probably emerged from the casual use of certain containers or from what a man or beast could lift. Most natural and convenient way of linear and area measurements in early days were provided by simple body measurements, viz; hands or feet.

Many early civilizations and ancient empires of the Near and Middle East are known to have devised units and standards of measurements, which found their way, mostly as a result of trade or conquest to western empires via Greek and Roman empires. Some of these further evolved in various systems of units and measurements of modern day. The important developments in the system of units and measurements are described in the following paragraphs.

1.2.1 Ancient Egyptian System

Devised about 3000 BC, the *Egyptian cubit* is generally recognised as having been the most common and wide spread standard of linear measurement in the world. The *cubit* was defined as the length of the arm from the elbow to the extended finger tips and a standard *cubit made of black granite* was kept by royal master, against which all the *cubit sticks* in use at that time in Egypt were calibrated at regular intervals.

The Egyptian cubit, which is equivalent to 20.62 inches or 524 millimetres, was subdivided into the basic subunit called *digit*. The *digit* was equivalent to a finger's breadth and 28 *digits* constituted a *cubit*. The subdivision of *cubit* was complicated by several other subunits in unusual way. Four *digits* equaled a *palm*, five a *hand*, twelve *digits*, or three *palms*, equaled a *small span*. Fourteen digits, or one-half a cubit, equaled a *large span*. Sixteen digits, or four palms, made one *t'ser*. Twenty digits, or five palms, were a *small cubit*.

The *digit* was further subdivided in an ingenious manner. The 14th digit on a cubit stick was subdivided into 16 equal parts. The next digit was subdivided into 15 parts, and so on, to the 28th digit which was divided into two equal parts. Thus, it was possible to make measurements to any digit fractions having denominator from 2 through 16. The smaliest division, 1/16 of a digit, was equivalent to 1/448 part of a cubit. Accuracy of the system is testified by the dimensions of the Great Pyramid of Gizeh. Although thousands of people were employed in construction of the pyramid, its side vary only within 0.05% from the mean length of 9,069.45 inches or 230.364 metres, i.e; about 4½ inches in 755 feet.

At a very early date, Egyptians developed methods and instruments for measuring and demarking land. The annual flood of the Nile created a need for surveying and demarking techniques so that property boundaries could be easily re-established after the flood receded.

The weight system of Egyptians appears to be founded on a basic unit called the *kite*, which varied from period to period over a long span of Egyptian history, ranging all the way from *4.5* to *29.9* grams. Weight system based on *kite* was, however, akin to modern decimal system, *10 kites* equalling *1 deben* and *10 debens* equalling *1 sep.*

For liquid measures, Egyptians used units called *rohin, hekat, khar* and *cubic cubit* which is equivalent to 0.14 cubic metre.

1.2.2 The Babylonian System

The Babylonian *mina* is possibly the earliest known weight unit. It is equivalent to 640 grams in one surviving form and equivalent to 978 grams in another. Weights of five *minas* in the shape of a duck and of thirty *minas* in the shape of swan have been found by archaeologists. The *shekel* which is familiar from the Bible as a standard Hebrew coin and weight is known to be of Babylonian origin. Most of the Babylonian weights and measures were used in carrying out commerce and trade throughout the Middle East and were gradually adopted by other countries at that time.

The *Kus* (equivalent to about 20.9 inches or 530 millimetres), also called *Babylonian Cubit*, was the basic Babylonian unit of length. The Babylonian *foot* was defined as 2/3 *kus* and the Babylonian *Shusi*, as 1/30 *kus* and was equal to 0.69 inch.

The Babylonian liquid measure, *ka* was defined as volume of a cube of one hand breadth (3.9 to 4 inches). The cube, however, had to contain a weight of one great *mina* of water. Incidently, the *ka* is very close to modern litre and great *mina* to modern kilogram. The *ka* was a subdivision of two other units; 300 *ka* equaled 60 *gin* and one *gur*. The *gur* was almost equal to 80 US gallons.

The Hittites, Assyrians, Phoenicians, and Hebrews systems were derived mostly from the Babylonian and Egyptian systems. Hebrews weight standards were based on the *mina*, the *talent* (a basic unit) and the *shekel*. The *sacred mina* equaled 60 *shekels* and the *sacred talent* equaled 3,000 *shekels* or 50 *sacred minas*. However, the *Talmudist mina* was equal to 25 *shekels* and *Talmudist talent* to 1500 *shekels* or 60 *Talmudist minas*.

The Hebrew standards of volume of liquid are not definitely known, the *bat* is considered to be equivalent to about 37 litres, the *hin* equaled slightly more than 6 litres and the *log* is slightly more than 0.5 litre. The close relationship between dry and liquid volumetric measures was the notable feature of the Hebrew System. The liquid *kor* was same size as the dry *homer* and liquid *bat* equaled the dry *'efa*.

1.2.3 Greeks and Romans System

Greeks derived their units of weights and measures partly from the Egyptians and partly from the Babylonians. By the first millennium BC, the commercial domination of the Mediterranean shifted to Greek and Romans. Effect of this is evident in the subsequent units of Greeks and Romans. The *finger* (19.3 mm) was a basic Greek unit of length; 16 *fingers* equaled one *foot* and 24 *fingers* equaled one *Olympic cubit*. A parallel to this is the Egyptian 24 *digits* equaling one *small cubit* which could be the result of commercial history of the era. A basic Greek unit of weight was the *talent* (equal to 25.8 kilograms) and *metretes* (equal to 39.4 litres) represented a measure of liquid.

The Romans adopted the Greek system and subdivided the *foot* into 12 *ounces*, or *inches* (*unciae*). A similar subdivision and same word, *unciae*, was used for one twelfth of a *pound* (*libra*) which equaled 327.45 grams. One *pace*, or *double step* equaled five *feet* and thousand *pace* made up the Roman *mile* (*mille passus*).

The basic Roman unit of volume was *sextarius* (0.53 litre) which had several subdivisions and multiples. The largest, the *amphora*, consisted 48 *sextarii* (25.5 litres or 6.7 US gallons).

1.2.4 Ancient Chinese Weights and Measures

Ancient Chinese system of weights and measures was totally different from Mediterranean-European systems, although, the principal features of western system could be seen. Like Mediterraneans, Chinese also used parts of the body as a source of units, for example, the distance from the pulse to the base of the thumb. The system was fundamentally chaotic as there existed no relationship between different types of units, such as those of length and those of volume. Units with the same name varied not only from region to region but also from trade to trade. The *mou*, unit of land measure, fluctuated from one-thirteenth of a hectare to one-eight of a hectare.

In 221 BC, the first emperor of China, Shih Huang Ti, attempted unification of units by adopting a regulation fixing the basic units. The basic unit of weight, the *shih* or *tan*, was fixed at about 60 kilograms, the two basic length measurements, the *chih* and the *chang*, were fixed at about 25 centimetres and 3 metres, respectively.

A unique characteristic of the Chinese system was its inclusion of an acoustical dimension. For example, a standard vessel for measuring grains and wine was defined not only as to weight but also as to pitch of the sound when struck. A vessel of uniform shape and fixed weight would give proper pitch only if it had a proper volume. This made the measuring instruments of a high order. Measures based on the length of a pitch pipe and its subdivision in terms of millet grains supplanted the measurements based on the human body.

1.2.5 Ancient Indian Weights and Measures

Attempts to standardize weights and measures in India can be traced back to the Indus Valley Civilization (3000-1500 BC). The stone weights found at Mohenjo-daro and Harappa were intended to be fractions and multiples of the *uncia* (1/8, 1/4, 1/2, 1 & 2 uncia). The relationship between *uncia, gram & grain* is : 10 *uncia* = 272 *gram* = 4200 *grain*. The *retti*, a half-black half-red seed of '*Abrus precatorius*' has served as measure of small weight (ref. Manusmriti, c.500 BC). The very small mote which is seen when the sun shines through a lattice, is called a *transarenu* (floating particle of dust) and 8 *transarenu* is equal to 1 *liksha* (egg of a louse), 3 *liksha* is equal to 1 *grain* of *black mustard*, 3 *grains* of *black mustard* is equal to 1 *grain* of *white mustard*, 6 *grains* of *white mustard* is equal to 1 *middle sized barley corn* and 3 *barley corn* is equal to 1 *krishnala* which is a *retti*. The mass rating of dust speck is of the order of 1/10 milligram. Manu's mass ratios were : 3200 *retti*=640 *masha*=40 *suvarna*=10 *pala*=1 *dharana* (which is 417.72 gram). The *maund* was used as a measure of weight in British India since early nineteenth century, it was divided into 40 *seers* = 3200 *tolas* = 576000 *grains* = 576000/7000 or $82^2/_7$ lb (avoirdupois).

The early record of measure of length is found in *Indus Inch* engraved on a broken piece of shell found at Mohenjo-daro in the Indus Valley, in the form of parallel lines 0.264 inch apart. A circle and a big dot at five spaces apart may be considered as *Indus Inch*=1.32 *inch*=2 *Sumerian shusi*. Evidence of survival of *Indus Inch* seems apparent in use of 33-inch north Indian *gaz* (yard) popular at the time of Emperor Akbar. Akbar's *yard*=33 inch=25 *Indus Inch* (*Illahi gaz*)=1/6 English *pole*.

The chief indigenous measure of area is the *biga* which is said to have been the area which a pair of oxen can plough in a day. The biga is defined as a sq. *jarib* which is equal to 60×60 sq. *gaz* where, *jarib*=20 *lath*=60 *gaz*=1440 *tasu*=5760 *pan*=10 *English pole*. Hence, a *biga*=100 sq. *pole*=*Roman jugerum*. Bhascara's *Lilavati* (c. A.D. 1150) has a reference of land-unit called *Nivaratna* which is equal to 20×20 bamboo poles of 10 cubits of 18 inches. Therefore, 1 *Nivaratna*=1 *myriad sq. yd.*=60^4 *sq. inch*.

The main Indian coinage was built around the *rupee* which is divided into 16 *annas* or 64 *pice* or 192 *pies*. In India, where the concept of and symbol for zero, and the decimal place-value notation for writing numbers had been invented, the merits of the metric system took a long time to be officially appreciated. Only in 1955, India formally decided to adopt it.

The Indian time measurement consists of dividing night and day into 60 parts, each called a *ghari*. A *ghari* is further divided into 60 parts, each being called a *pal*. The night and day is also divided into 4 parts each called *pahr*.

The Indian prefix system was of high reckoning, 100,000 being called a *lak*, 100 *laks* make a *krur* (*crore*), 100 *krurs* a *arab*, 100 *arabs* a *krab* (*kharab*), 100 *krab* a *nil*, 100 *nil* a *padam* and 100 *padam* a *sang*.

1.2.6 Medieval Weights and Measures

Roman system with its Greek, Babylonian and Egyptian roots formed the basis of Medieval European system of weight and measure which had large number of regional and national variants dictated by the needs of the commercial activities of the era.

As the commercial interactions amongst ratios increased, need for uniformity was felt. But, the attempt to bring uniformity by Charlemagne in 9th Century was in vain. However, the great trade fairs of 12th and 13th Centuries, such as the Champagne Fairs, rigidly enforced uniformity on merchants of all nationalities within fair grounds and this had some effect on efforts towards standardization. A universal measure of wool cloth, the *ell*, is a good example. The *ell* of champagne, equaled *two feet six inches* measured against an iron standard kept with the Fair Keepers, was accepted by Ypres, Ghent, Arras and several other great cloth-manufacturing cities of northwest Europe, even though their *bolts* varied in length. However, the *ell* itself varied in several other parts of Europe.

The *libra*, the Roman unit of weight acquired different names in parts of northern Europe but retained its identity in the English abbreviation of *pound* as *lb*. The Roman *mile* remained in Medieval

Europe, while the *pace* on which it was based vanished. Instead of thousand *pace*, now *mile* consisted varying number of *feet* and *yards*. The *pinte* or *pint* was the unit of liquid measure which approximately equaled the modern English *quart*.

1.2.7 The English System

By the time of Magna Carta (1215), the problem of variation in weights and measures became so common that a clause was to be inserted in the charter to correct those on grain and wine. A royal ordinance entitled "Assize of weights and Measures", was issued few years later. This was a breakthrough in defining a broad list of units and standards, so successfully that it remained in force for nearly 600 years. A *standard yard* divided into the traditional three feet, each of 12 inches, was defined as exactly equal to "the standard iron yard of Lord the King". The *perch* (later *rod*) was defined as 5½ yards or 16½ feet. The *inch* was subdivided into three barley corns. The *furlong* was standardized as an eight of a *mile*, the acre equal to an area 4 rods wide by 40 rods long.

The Champagne fairs had an influence on separate English pounds for troy weight used for weighing gold and silver bullion (the name probably from Troyes—a principal fair city of Medieval time) and avoirdupois—used for goods that had to be weighed, e.g; sugar, salt, alum, dyes, grain. The troy pound and apothecaries pound for weighing drugs contain only 12 troy ounce. The *stone*, a multiple of English pound equaled 14 pounds and 26 *stones* or 364 pounds made a *sack* of raw wool.

The bronze or brass standards were sent out to provincial towns from London in order to ensure compliance. But, discrepancies crept in. Therefore, in 1496 and then in 1588 under Elizabeth-I, new standards were made and sent out. During next 200 years, no revision of law was felt necessary, but several refinements and redefinitions were added.

Edmund Gunter, a 17th century mathematician conceived the idea of a chain equal to acre's breadth (40 perches or rods) and divided it into 100 *links*. In 1701, a dry measure, the *corn bushel*, was defined as "any round measure with a plain and even bottom, being 18½ inches wide throughout and 8 inches deep". Similarly, in 1707, a *wine gallon* was defined as a round measure with a even bottom and containing 231 cubic inches; however, the *ale gallon* was retained at 282 cubic inches. There was also a *corn gallon* and an older, slightly smaller *wine gallon.*

A new Weights and Measures Act of 1824 attempted to clear some of the medieval confusion and stressed on the accuracy of standards. A single gallon was defined as the "volume occupied by 10 imperial pounds of distilled water weight in air against brass weights with the water and the air at a temperature of 62 degrees of Fahrenheit's thermometer and with the barometer at 30 inches".

In the Act of 1878, gallon remained same but yard was redefined as "the straight line or distance between the centers of two gold plugs or pins in the bronze bar ...measured when the bar is at the temperature of 62 degrees of Fahrenheit's thermometer, when it is supported by bronze rollers placed under it in such a manner as best to avoid flexure of the bar...". Thereafter, in 1963, an Act of Parliament redefined weights and measures in terms of *metric system*, with a national changeover beginning two years later.

1.2.8 The US Customary System

In US, George Washington stressed on the need for uniformity in currency, weights and measures in his first message to Congress in 1790. The decimal form was accepted for currency but not for weights and measures. The vast inertia of the English weights and measures system used in industry and commerce; and tools and machines based on the same coupled with popularity and psychology acted as deterrent to any change. However, the metric system was becoming a reality in France, and in 1821, Secretary of State, John Quincy Adams reported to the Congress that the metric system is "worthy of acceptance....... beyond a question". But Adams admitted the impossibility of winning acceptance for it in US, until a future time.

Instead of adopting metric system, U.S. tried to bring its system into closer harmony with the English system, removing the deviations, such as, "Queen Anne's gallon" of 231 cubic inches which the U.S.

still used, whereas, British had discarded the same in 1824. Construction of standards was undertaken by the office of Standard Weights and Measures under the Treasury Department. The standard for the yard was the one imported from London which guaranteed a close harmony with English yard; but Queen Anne's gallon was retained. The avoirdupois pound (7,000 grains) and the troy pound (5,760 grains) was exactly same as the British but the U.S. bushel (2,150.42 cubic inches) deviated from the British. The U.S. bushel was derived from the "Winchester bushel", a surviving standard dating back to the fifteenth century which was replaced by the British Act of 1824. Thus, the U.S. gallon and bushel remained more truly medieval than their British counter parts.

The office of Standard Weights and Measures was upgraded, by an Act of Congress effective from July I, 1901 to the National Bureau of Standards. Besides, the construction of physical standards and coordination in establishment of standard practices, its enlarged role and functions included activities such as; developing methods for testing materials and structures, carrying out research in engineering; physical science and mathematics; compilation and publication of general scientific and technical data etc.

The new Bureau sponsored a National Conference on weights and measures; one of the principal functions of the annual conference became the updating of a model state law on weight and measure which brought about virtual uniformity in legislation among the states. However, U.S. remained unique among major nations in refraining from exerting control at national level except for the Metric Act of 1866 which permitted the use of the metric system.

1.2.9 The Metric System

One of the most significant achievements of the French Revolution was the establishment of the metric system of weights and measures. The scientific growth and need for effective scientific communication had led the European scientists to discuss the desirability of a new, rational and uniform system to replace the national and regional variants. The proposal made as early as 1670 became the basis of eventual development of metric system.

Gabriel Mouton, the vicar of St. Paul's Church in Lyon, suggested a length measure based on the arc of one minute of longitude subdivided decimally. Three of the major characteristics of the metric system contained in Mouton's proposal are (i) decimalization (ii) rational prefixes and (iii) the earth's measurement as basis for a definition.

It took 120 years of discussions, amendments and criticism before the fall of the Bastille and the creation of the National Assembly which made the metric system a political possibility. In April of 1790, on the initiative of Tolerand, one of the foremost members of the Assembly, a directive was issued to the French Academy of Sciences to prepare a report. The Academy ultimately recommended that the "length of the meridian passing through Paris be determined from the North Pole to the Equator, and that one ten-millionth of this distance be termed the metre and form the basis of a new decimal linear system; and further, that a new unit of weight should be derived from the weight of a cubic metre of water. A list of prefixes for decimal multiples and submultiple was also proposed. The National Assembly endorsed the report and directed that the necessary meridional measurements be taken.

A Committee of 12 mathematicians, geodesics, and physicists met with Louis XVI, who gave his formal approval on June 19, 1791. Later, two engineers, Jean Delambre and Pierre Machain undertook the arduous task of measuring the distance on the meridian. In 1793, a provisional metre was constructed from geodetic data already available. In 1795, the firm decision was taken to enact adoption of the metric system for France. With formal presentation to the assembly of the standard metre, as determined by Delambre and Machain, the metric system became a reality in June 1799. The motto adopted was "For all people, for all time".

The standard *metre* was defined as one ten millionth part of a meridional quadrant of the earth as measured by Delambre and Machain. The *gram*, the basic unit of mass, was made equal to the mass of a cubic centimetre of pure water at the temperature of its maximum density (4°C; 39.2°F). A platinum cylinder known as the kilogram of the Archives was fabricated and declared the standard for 1,000 grams. The *litre* was defined as the volume equivalent to the volume of a cube having each side equal to ten centimetres or one decimetre.

The *are* was defined as the measure of area equal to a square of ten metre side. The *hectare*, used in practice is a multiple of *are* equivalent to 100 *are*. The *stere* was defined as the unit of volume equivalent to a cubic metre. The names of multiples and sub multiples of all units were also standardized based on Greek prefixes.

The successes of French Revolution contributed to the conquest of the metric system in Europe but it required a long time to overcome the inertia of customary system. Even in France, Napoleon had to issue a decree permitting use of the old medieval system along with the new metric system. But, the advantages of the metric system vis-a-vis customary system proved decisive and in 1840, it was established as a legal monopoly in France. Thereafter, the progress of the metric system has been steady, in many cases the metric system was adopted during course of a political upheavals, e.g; Latin America, the U.S.S.R., China and in Japan with the Meiji restoration of 1868.

However, in Britain, the commonwealth nations still retained the British Imperial System and in U.S., there was no major progress of metric system. The U.S. signed the Metric Convention of 1875 and later announced that International Prototype *metre* and the *kilogram* would be regarded as fundamental standards.

The 20th Century saw spontaneous increase in the use of metric system in various segments of commerce and industry. In Britain and the U.S., it became almost universally employed in the scientific and medical professions. Industries related to chemicals, electric power, photography, optometry and electronics adopted the metric system atleast in part.

The U.S. army also adopted metric for ordnance. Legislative proposals in favour of metric began to be made in British Parliament and U.S. Congress. The Congress passed a legislation in 1968 to investigate, research and survey the impact made on the U.S. due to the worldwide use of the metric system. The Parliament went a step ahead and took the steps to begin a long-range program of change over.

In the meantime, the metric system itself was replaced by the new System Internationale (SI), which is basically an expansion of the 18th century metric to accommodate the scientific and technological developments of the 20th century.

2

SI Units (Systéme International D' Unites)

2.1 Introduction

The new International System of Units (SI) grew out of the problems faced by growing scientific community in meaningful comparisons and communication of results of scientific inventions. The proliferation of existing systems which varied greatly was causing several problems. The original 18th century standards were not accurate enough to the degree demanded by 20th century scientific requirements. Thus, new definitions and a more rationalised system was the need of the day. The new Systéme International d'Unites (SI) was thus formulated in October 1960, in the 11th General Conference on Weight and Measures meeting at Paris, the birthplace of the metric system. The SI is rationalization of the metric system and it provides a coherent and elegant framework which has removed many of the artificial divisions between quantities in, e.g., mechanics, electrics and heat. Consequently, it has found wide acceptance by scientific community all over the world and has been universally adopted.

2.2 SI Base Units

The SI consists of seven base units, two supplementary units and a series of derived units consistent with the base and supplementary units.

2.2.1 Base Units

	Measurement	*Unit*	*Symbol*	*Definition*
1.	Length	metre	m	The metre is the length equal to 1 650 763.73 wavelengths in vacuum of the radiation corresponding to the transition between the levels $2p_{10}$ and $5d_5$ of the krypton-86 atom (11th CGPM*, 1960; resolution 6).
2.	Time	second	s	The second is the duration of 9 192 631 770 periods of the radiation corresponding to the transition between the two hyperfine levels of the ground state of the caesium-133 atom (13th CGPM*, 1969; resolution 1).
3.	Mass	kilogram	kg	The kilogram is equal to the mass of platinum-iridium prototype kept at the International Bureau of Weights and Measures. This is the only base unit defined by an artifact (1st CGPM*, 1889 and 3rd CGPM, 1901).
4.	Thermo-dynamic temperature	kelvin	K	The kelvin, used to express an interval or a difference of temperature, is the fraction 1/273.16 of the thermodynamic temperature of the triple point of water (13th CGPM*, 1967; resolution 4).

5.	Electric current	ampere	A	The ampere is that constant current which, if maintained in two straight parallel conductors of infinite length, of negligible cross-section, and placed 1 metre apart in vacuum, would produce between these conductors a force equal to 2×10^{-7} newton per metre of length (CIPM**, 1946; resolution 2 approved by the 9th CGPM*, 1948).
6.	Luminous intensity	candela (kan-dell'-ar)+	cd	The candela is the luminous intensity, in the perpendi cular direction, of a surface of 1/600 000 square metre of a black body at the temperature of freezing platinum (2042 K) under a pressure of 101 325 newton per square metre (13th CGPM*, 1967; resolution 5).
7.	Amount of a substance	mole	mol	The mole is the amount of substance of a system which contains as many elementary entities as there are atoms in 0.012 kilogram of carbon-12 (13th CGPM+, 1967; resolution 5).

* *Conférence Générale des Poids et Mesures*

** *Comité International des Poids et Mesures*

\+ *Latin for 'candle'*

2.2.2 Supplementary Units

	Measurement	*Unit*	*Symbol*	*Definition*
1.	Plane angle	radian	rad*	The plane angle subtended at the center by an arc of unit length at unit radius.
2.	Solid angle	steradian	sr**	The solid angle subtended at the center by unit area of a spherical surface at unit radius.

* *equivalent combination of other units : m/m (i.e; length of circular arc ÷ radius)*

** *equivalent combination of other units : m^2/m^2 (i.e; area of spherical surface ÷ $radius^2$)*

2.2.3 Important Derived Units

Derived SI units, as shown below, are coherent, i.e; unit value of a derived quantity is obtained by multiplication and division of unit values of basic quantities. Relationships of the derived SI units are depicted in Fig.1.

Quantity	*Unit*	*SI Symbol*	*Formula*
Absorbed dose	gray	Gy	J/kg
Acceleration	metre/second square		m/s^2
Activity (of a radio nuclide)	becquerel	Bq	1/s
Area	square metre		m^2
Celsius temperature	degree Celsius	^0C	^0C=K-273.15
Dose equivalent	sievert	Sr	J/kg
Electric charge (quantity)	coulomb	C	A.s
Electrical capacitance	farad	F	A.s/V
Electrical conductance	siemens	S	A/V
Electrical inductance	henry	H	V.s/A
Electrical potential difference (electromotive force)	volt	V	W/A
Electrical resistance	ohm	Ω	V/A

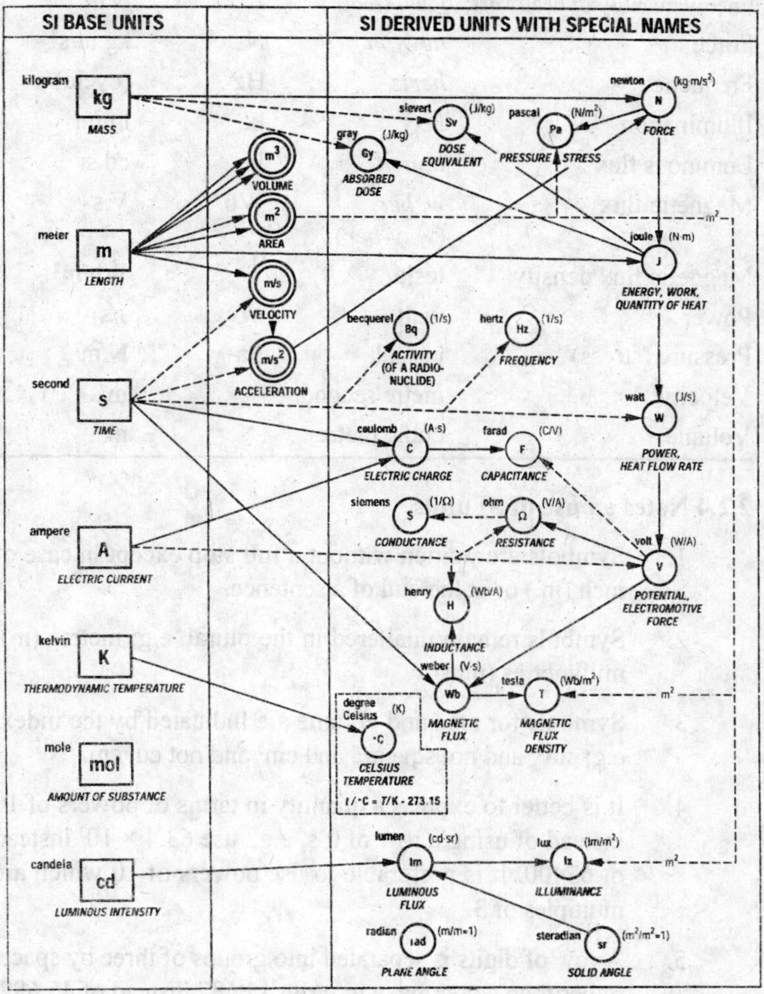

Figure 1. Relationships of the SI Base Units and the SI Derived Units with special names

(solid lines indicate multiplication & broken lines indicate division)

Energy/quantity of heat/work	joule (*jool*)	J	N.m
Force	*newton*	N	kg.m/s^2
Frequency	*hertz*	Hz	(cycle)/s
Illumination	*lux*	lx	lm/m^2
Luminous flux	*lumen*	lm	cd.sr
Magnetic flux	*weber,* (*way-ber*)	Wb	V.s
Magnetic flux density	tesla	T	Wb/m^2
Power	watt	W	J/s
Pressure (stress)	pascal	Pa	N/m^2
Velocity	metre/second		m/s
Volume	cubic metre		m^3

2.2.4 Notes on use of SI units

1. Symbols are written without a full stop except in case of inch (in.) or at the end of a sentence.
2. Symbols remain unaltered in the plural, e.g; metres (m), millimetres (mm).
3. Symbols for area and volume are indicated by the index, e.g; cm^2, and not sq. cm., and cm^3 and not cu. cm.
4. It is better to express a quantity in terms of powers of 10 instead of using a row of 0's, e.g., use 63.4×10^3 instead of 63400. It is preferable to use powers of 10 which are multiples of 3.
5. A row of digits is separated into groups of three by space, rather than a comma, e.g; write 45 687 instead of 45,687. If the figure happens to consist of four digits only, it is left without separation, e.g; 4576 will be written without a gap between 4 and 5.
6. The product of two or more units in symbolic form is preferably indicated by a dot midway to avoid risk of confusion with another unit symbol, e.g; use N.m or N m, but not Nm.

2.3 Non-SI Units, the use of which are likely to continue

Quantity	*Name of Unit*	*Unit Symbol*	*Definition*
Time	minute	min	1 min = 60 s
	hour	h	1 h = 60 min
	day	d	1 day = 24 h
Plane angle	degree	^{0}	$1^0 = \pi/180$ rad
	minute	'	$1' = 1/60^0$
	second	"	$1'' = 1/60'$
Volume	litre	l	1 l $=(dm)^3=(1000\ cm^3)$
Mass	tonne	t	1 t = 1000 kg
Fluid Pressure	bar	bar	1 bar = 10^5 Pa
Land area	hectare	ha	1 ha=10,000 m^2
Speed	kilometre/hour	$km.\ h^{-1}$	
Angular speed	revolution/second	$rev.s^{-1}$	

2.4 SI Unit Prefixes

Multiples and submultiples	*Prefix*	*SI Symbol*
10^{18}	exa	E
10^{15}	peta	P
10^{12}	tera	T
10^9	giga (*jy'-ga*)	G
10^6	mega (*meg'-ga*)	M
10^3	kilo	k
10^2	hecto	h
10^1	deka	da
10^{-1}	deci	d
10^{-2}	centi	c
10^{-3}	milli	m*
10^{-6}	micro	µ
10^{-9}	nano	n
10^{-12}	pico (*pee'-ko*)	p
10^{-15}	femto	f
10^{-18}	atto	a

** The symbol 'm' when used alone represents metre.*

2.5 Symbols and Other Abbreviations

Unit	*Symbol*
Angstrom	Å
Apostilb	asb
Are	a
Atmosphere (standard)	atm
Atmosphere (technical)	at
Atomic mass unit	amu
Bar	bar
Barn	b
Bel	B
Bit	bit
Bohr magneton	μ_B
British thermal unit	Btu
Calorie	cal
Candle	c
Cent	c
Centimetre	cm
Cubic x	x^3
Curie	c
Cycle	c
Day	d
Debye	D
Decibel	dB
Degree	^{0} (deg)
Dioptre	D
Dyne	dyn
Einstein	E
Electron-volt	ev
Erg	erg
Feet/foot	ft
Fermi	f
Gal	G
Gallon	gal
Gallons per minute	gpm
Gamma	g
Gauss	Gs
Gilbert	Gb
Gram	g
Hectare	ha
Horsepower	hp
Hour	h
Hundred weight	cwt
Inch	in.
Knot	kn
Lambert	L
Litre	l
Maxwell	Mx
Micron (micrometre)	$\mu(\mu m)$
Mile (nautical)	n.mile
Millimetre of mercury	mm Hg
Millimetre of water	mm H_2O
Minute	min
Minute(angular)	'
Mole	mole
Neper	n
Nit	nt
Octave	oct
Oersted	Oe
Ounce	oz

Unit	*Symbol*	*Unit*	*Symbol*
Parsec	pc	Rutherford	Rd
Phon	P	Rydberg	Ry
Phot	ph	Savart	Sav
Pieze	pz	Second, angular	″
Pint	pt	Square x	x^2
Poise	P	Sthene	sn
Pound	lb	Stilb	Sb
Pound force	lbf	Stokes	St
Poundal	pd	Ton	ton
Quart	qt	Tonne	t
Quintal	q	Torr	torr
Rad	rad	X unit	XU
Revolution	r	Yard	yd
Revolutions per minute	rpm	Year	y
Roentgen	r		

2.6 Some Units and Names of Units Having Limited Use or Not Introduced Officially

Unit		*Description*
Acohm	:	acoustic ohm - cgs unit of acoustic resistance
Biot	:	cgsm unit of current intensity, equal to 10A
Eötvös	:	unit of the gradient of the force of gravity, equal to a change in the acceleration of gravity by 1 cm/s^2 per centimetre
Franklin	:	cgs unit of charge, equal to 3.34×10^{10}C
Gal	:	unit of acceleration, equal to 1 cm/s^2
Inerta	:	unit of mass in the mks (force) system, equal to 9.81 kg
Kayser	:	unit of wave number, equal to cm^{-1}

Kilopond	:	name of kilogram - force used in German literature
Lenz	:	unit of magnetic field intensity, equal to 1 A/m
Magn	:	unit of "absolute" permeability, equal to $10^7/4\pi f/m$
Mes	:	unit of velocity, equal to 1 m/s
Mho	:	unit of conductivity, the same as siemens
Point (*of the compass*)	:	a unit of angle used in navigation, equal to 1/32 of a circle, i.e.; 11.25^0
Rhe	:	cgs unit of fluidity, the fluidity of a liquid having a viscosity of 1 poise.

2.7 Preferred Scales for Drawings

Type of drawing		*Scale*
Location drawings		
	block plan	1 : 2500
		1 : 1250
	site plan	1 : 500
		1 : 200
	location plan	1 : 100
Detail drawings		
	control detail	1 : 50
	working detail	1 : 20
		1 : 10
Fabrication drawings		
	detail	1 : 5
	template	1 : 1

Note : Drawings are scaled in metre (m) and millimetre (mm) as 0.000 m.

2.8.1 American Steel and Wire Gage (Washburn and Moen)

(Decimal Sizes and Feet per Pound)

AS & W steel wire gage no.	Deci-mally, in.	Ft per lb	AS & W steel wire gage no.	Deci-mally, in.	Ft per lb
1—	0.283	4.68	8—	0.162	14.3
¼	0.278	4.85	¼	0.159	14.8
½	0.273	5.03	½	0.155	15.6
¾	0.268	5.22	¾	0.152	16.2
2—	0.2625	5.44	9—	0.1483	17.1
¼	0.258	5.63	¼	0.143	17.8
½	0.253	5.86	½	0.142	18.6
¾	0.248	6.10	¾	0.138	19.7
3—	0.2437	6.31	10—	0.135	20.6
¼	0.239	6.56	¼	0.131	21.6
½	0.235	6.79	½	0.128	22.9
¾	0.230	7.09	¾	0.124	24.4
4—	0.2253	7.39	11—	0.1205	25.8
¼	0.221	7.68	¼	0.117	27.4
½	0.216	8.04	½	0.113	29.4
¾	0.212	8.34	¾	0.109	31.6
5—	0.207	8.71	12—	0.1055	33.7
¼	0.203	9.10	¼	0.102	36.1
½	0.200	9.37	½	0.099	38.3
¾	0.196	9.76	¾	0.095	41.5
6—	0.192	10.2	13—	0.0915	44.8
¼	0.188	10.6	¼	0.089	47.3
½	0.185	11.0	½	0.086	50.7
¾	0.181	11.4	¾	0.083	54.4
7—	0.177	12.0	14—	0.080	58.6
¼	0.173	12.5	¼	0.078	61.6
½	0.170	13.0	½	0.076	64.9
¾	0.166	13.6	¾	0.074	68.5

AS & W steel wire gage no.	Decimally, in.	Ft per lb	AS & W steel wire gage no.	Decimally, in.	Ft per lb
15—	0.072	72.3	23—	0.0258	563.3
¼	0.070	76.5	¼	0.0251	595.1
½	0.067	83.5	½	0.0244	629.7
¾	0.065	88.7	¾	0.0237	667.5
16—	0.0625	96.0	24—	0.0230	708.7
¼	0.060	104.2	¼	0.0224	747.2
½	0.058	111.5	½	0.0217	796.2
¾	0.056	119.6	¾	0.0211	842.1
17—	0.054	128.6	25—	0.0204	901
¼	0.052	138.7	¼	0.0198	956
½	0.051	144.2	½	0.0193	1007
¾	0.0491	155.5	¾	0.0187	1072
18—	0.0475	166.2	26—	0.0181	1144
¼	0.0459	177.9	¼	0.0179	1170
½	0.0443	191.1	½	0.0177	1197
¾	0.0426	206.6	¾	0.0175	1224
19—	0.0410	223.0	27—	0.0173	1253
¼	0.0394	241.6	¼	0.0170	1297
½	0.0379	261.1	½	0.0168	1329
¾	0.0363	284.5	¾	0.0165	1377
20—	0.0348	300.0	28—	0.0162	1429
¼	0.0340	324.3	¼	0.0159	1483
½	0.0332	340.2	½	0.0156	1540
¾	0.0325	355.1	¾	0.0153	1602
21—	0.0317	373.1	29—	0.0150	1666
¼	0.0309	392.7	¼	0.0148	1712
½	0.0301	413.8	½	0.0145	1783
¾	0.0294	433.7	¾	0.0143	1833
22—	0.0286	458.4	30—	0.0140	1913
¼	0.0279	481.7	¼	0.0138	1969
½	0.0272	506.8	½	0.0136	2027
¾	0.0265	533.9	¾	0.0134	2088

AS & W steel wire gage no.	Deci-mally, in.	Ft per lb	AS & W steel wire gage no.	Deci-mally, in.	Ft per lb
31—	0.0132	2152	38—	0.0080	7858
¼	0.0131	2185	¼	0.0079	6007
½	0.0130	2218	½	0.0078	6163
¾	0.0129	2253	¾	0.0076	6491
32—	0.0128	2288	39—	0.0075	6665
¼	0.0126	2361	¼	0.0074	6846
½	0.0123	2478	½	0.0073	7037
¾	0.0121	2561	¾	0.0071	7437
33—	0.0118	2693	40—	0.0070	7652
¼	0.0115	2834	¼	0.0069	7874
½	0.0111	3043	½	0.0068	8110
¾	0.0108	3215	¾	0.0067	8352
34—	0.0104	3466	41—	0.0066	8607
¼	0.0102	3605	42—	0.0062	9753
½	0.0100	3749	43—	0.0060	10415
¾	0.0097	3985	44—	0.0058	11145
35—	0.0097	4194	45—	0.0055	12394
¼	0.0094	4243	46—	0.0052	13866
½	0.0093	4335	47—	0.0050	14997
¾	0.0091	4527	48—	0.0048	16273
36—	0.0090	4629	49—	0.0046	17718
¼	0.0089	4733	50—	0.0044	19366
½	0.0087	4953			
¾	0.0086	5069			
37—	0.0085	5189			
¼	0.0084	5313			
½	0.0083	5443			
¾	0.0081	5714			

2.8.2 Relationship of Gage Numbers in Common Use

	Weight gages							Thickness gages			
Name of gage...	United States standard, USSG		Galvanized sheet, GSG		Tin-plate, TPG			Steel wire Washburn & Moen or W & M wire, U.S. steel, Steel	Music-wire, MWG	Brown & Sharpe, B & SG, AWG	Stubs iron wire, WWG. BWG
Principal use...	Uncoated carbon-steel sheets and light plates		Galvanized sheet steel		Tin plate			Steel wire, except music wire	Steel music wire	Non-ferrous sheets and wire	Flats, plates, and wire
Gage no.	Equivalent thickness, in.	Psf	Psf	Oz per sq ft.	Psf	Lb per base box	Symbol	Thickness, in.	Thickness, in.	Thickness, in.	Thickness, in.
1	2	3	4	5	6	7	8	9	10	11	12
7/0	0.4902	20.0000						0.4900			
6/0	0.4596	18.7500						0.4615	0.004	0.5800	
5/0	0.4289	17.5000						0.4305	0.005	0.5165	0.500
4/0	0.3983	16.2500						0.3938	0.006	0.4600	0.454

1	2	3	4	5	6	7	8	9	10	11	12
3/0	0.3676	15.0000						0.3625	0.007	0.4096	0.425
2/0	0.3370	13.7500						0.3310	0.008	0.3648	0.380
0	0.3064	12.5000						0.3065	0.009	0.3249	0.340
1	0.2757	11.2500						0.2830	0.010	0.2893	0.300
2	0.2604	10.6250						0.2625	0.011	0.2576	0.284
3	0.2451	10.0000						0.2437	0.012	0.2294	0.259
4	0.2298	9.3750						0.2253	0.013	0.2043	0.238
5	0.2145	8.7500						0.2070*	0.014	0.1819	0.220
6	0.1991	8.1250						0.1920*	0.016	0.1620	0.203
7	0.1838	7.5000						0.1770*	0.018	0.1443	0.180
8	0.1685	6.8750	7.0312	112.5				0.1620*	0.020	0.1285	0.165
9	0.1532	6.2500	6.4062	102.5				0.1483*	0.022	0.1144	0.148
10	0.1379	5.6250	5.7812	92.5				0.1350*	0.024	0.1019	0.134
11	0.1225	5.0000	5.1562	82.5				0.1205*	0.026	0.0907	0.120
12	0.1072	4.3750	4.5312	72.5				0.1055*	0.029	0.0808	0.109
13	0.0919	3.7500	3.9062	62.5				0.0915*	0.031	0.0720	0.095
14	0.0766	3.1250	3.2812	52.5				0.0800*	0.033	0.0641	0.083
15	0.0689	2.8125	2.9687	47.5				0.0720*	0.035	0.0571	0.072
16	0.0613	2.5000	2.6562	42.5				0.0625*	0.037	0.0508	0.065
17	0.0551	2.2500	2.4062	38.5				0.0540*	0.039	0.0453	0.058
18	0.0490	2.0000	2.1562	34.5				0.0475*	0.041	0.0403	0.049

1	2	3	4	5	6	7	8	9	10	11	12
19	0.0429	1.7500	1.9062	30.5				0.0410*	0.043	0.0359	0.042
20	0.0368	1.5000	1.6562	26.5				0.0348	0.045	0.0320	0.035
21	0.0337	1.3750	1.5312	24.5				0.0317	0.047	0.0285	0.032
22	0.0306	1.2500	1.4062	22.5				0.0286	0.049	0.0253	0.028
23	0.0276	1.1250	1.2812	20.5	1.1250			0.0258	0.051	0.0226	0.025
					1.079	235	6X				
					1.047	228	6XL				
24	0.0245	1.0000	1.1562	18.5	1.0000			0.0230	0.055	0.0201	0.022
					0.987	215	5X				
					0.964	210	D2X				
					0.955	208	5XL				
					0.895	195	4X				
25	0.0214	0.8750	1.0312	16.5	0.8750			0.0204	0.059	0.0179	0.020
					0.863	188	4XL				
					0.827	180	DX				
					0.804	175	3X				
					0.771	168	3XL				
26	0.0184	0.7500	0.9062	14.5	0.7500			0.0181	0.063	0.0159	0.018
					0.748	163					
					0.712	155	2X				

* Three intermediate fractional gages sometimes used are omitted in this table.

2.8.3. Standard wire gage

Gage No.	*in.*	*mm*	*Gage No.*	*in.*	*mm*
0000	.400	10.160	24	.022	.559
000	.372	9.0449	25	.020	.508
00	.348	8.839	26	.018	.457
0	.324	8.230	27	.0164	.4166
1	.300	7.620	28	.0148	.3759
2	.276	7.010	29	.0136	.3454
3	.252	6.401	30	.0124	.3150
4	.232	5.893	31	.0116	.2946
5	.242	5.385	32	.0108	.2743
6	.182	4.877	33	.0100	.2540
7	.176	4.470	34	.0092	.2337
8	.160	4.084	35	.0084	.2134
9	.144	3.658	36	.0076	.1930
10.	.128	3.251	37	.0068	.1727
11.	.116	2.946	38	.0060	.1524
12.	.104	2.642	39	.0052	.1312
13.	.092	3.337	40	.0048	.1219
14.	.080	2.032	41	.0044	.1118
15.	.072	1.829	42	.0040	.1016
16.	.064	1.626	43	.0026	.0914
17.	.056	1.422	44	.0032	.0813
18.	.048	1.219	45	.0028	.0711
19.	.040	1.016	46	.0024	.0610
20.	.036	.914	47	.0020	.0508
21.	.032	.813	48	.0016	.0408
22.	.028	.711	49	.0012	.0305
23.	.024	.610	50	.0010	.0254

2.9 Some Definitions of Selected Quantities Commonly Used and Their Units

2.9.1 Mechanical Units

Capacity: The fundamental units of capacity in metric and British systems are litre and gallon, respectively. They are defined in terms of the volume occupied by a definite quantity of water, specified either by its mass or by its weight in air as follows :

Litre: Volume occupied by a mass of one kilogram of pure air-free water at the temperature of its maximum density (4^{0}C) and under normal atmospheric pressure (101 325 newton per square metre) [Standards of Weights and Measures Act (India), 1956].

Gallon: Volume occupied by a quantity of distilled water which weighs 10 pounds in air against brass weights, with the water and air at the temperature of 62^{0}F and with the barometer at 30 inches [Weights and Measures Act (UK), 1978].

In USA, two types of units of capacity are used, one for liquids and the other for dry goods. The fundamental units for the two types are; *gallon*—for liquids and *bushel*—for dry goods. They are defined as; 1 gallon (US) = 231 cubic inches and 1 bushel (US) = 2150.42 cubic inches.

The principal conversion factors for inter-conversion of cubic measures and capacities are; 1 gallon (UK) = 4.545 96 litres and 1 litre = 1000.028 cubic centimetres.

Area: Bigha *:* Bigha is unit of land measurement which is equivalent to five eighth of an acre. The *bigha* is equal to a sq. *jarib* = 60×60 sq. *gaz* (33 inch)=100 sq. *Pole=Roman jugerum.*

Fluid flow: **Miners' inch**: A miner's inch is a measure for flow of water and is the quantity of water that will flow in one minute through an opening 1 inch square in a plank 2 inches thick under a head of 6½ inches to the center of the orifice. This is equivalent, approximately, to 1.53 cubic feet, or 11½ gallons per minute.

Power: Metric Horsepower: The unit of power adopted for engineering work, as defined in the English system of measurement, is equal to 33,000 foot-pounds per minute or 550 foot-pounds per second. The metric horsepower (also referred as Pferde Starke or PS), used in countries where the metric system is employed, is equal to 75 kilogram-metres per second, which is only 32,550 foot-pounds per minute or 542.5 foot-pounds per second. Hence, it is evident that the metric horsepower is about 1.5 per cent smaller than the horsepower employed in the countries using the English system of measurements.

2.9.2 Electrical Units

Charge: The unit of charge (quantity) is the *coulomb* (C), defined as the quantity of electricity transported per second by a current of 1 ampere.

Potential Difference: The unit of potential difference is the *volt,* (V) defined as that difference of electrical potential between two points of a wire carrying a constant current of 1 ampere when the power dissipation between those points is 1 watt.

Resistance: The unit of resistance is the *ohm,* (Ω) defined as the electrical resistance between two points of a conductor when a constant potential difference of 1 volt applied between these points produces in the conductor a current of 1 ampere.

Conductance: The unit of conductance is the *siemens,* (S) defined as the electrical conductance between two points of a conductor when a constant potential difference of 1 volt applied between these points produces in the conductor a current of 1 ampere.

Inductance: The unit of inductance is the *henry,* (H) defined as the inductance of a closed circuit in which an electromotive force of 1 volt is produced when the current in the circuit varies uniformly at the rate of 1 ampere per second.

Capacitance: The unit of capacitance is the *farad,* (F) defined as the capacitance of a capacitor between the plates of which there appears a difference of potential of 1 volt when it is charged by 1 coulomb.

Electrical Intensity: The electrical intensity at a point in an electric field is the force exerted on unit charge (1 coulomb) placed at that point, assuming that the field is not disturbed by doing so. Unit *voltmetre*$^{-1}$ (which is equivalent to the newton coulomb^{-1}).

Coulomb's Electrostatic Law: Appears in the form $F = Q_1Q_2 / 4\,\varepsilon_0\varepsilon_r\pi d^2$, where Q_1 and Q_2 are the two charges situated a distance *d* apart in a medium whose permittivity relative to that of free space is ε_r. The permittivity of free space $\varepsilon_0 \simeq (1/36\pi) \times 10^{-9}$ farad/metre. ε_r is a pure number. The product $\varepsilon_0\varepsilon_r$ is analogous to the dielectric constant as defined in the cgs system.

2.9.3 Magnetic Units

Magnetic Intensity (H): is defined through Ampere's theorem for the intensity due to a current element. In the usual notation, $H = I.ds.\sin\theta/4\pi r^2$. Unit, *ampere metre*$^{-1}$.

Magenetic Flux (Φ) of the induction B: is defined as $\int B.n\,dA$ where, *n* is a unit vector perpendicular to an element of area dA.Unit, *weber* (Wb).

Magenetic Flux Density or Induction(B): is defined through the equation for the force on a current element placed in a magnetic field, viz; $F= B.I.ds.\sin\theta$, in the usual notation, Unit, *Tesla* (T). $B=\mu_o\mu_r H$ where, μ_r is the relative permeability of the medium with respect to free space and μ_o is the permeability of free space. $\mu_o=4\pi x 10^{-7}$ henery metre^{-1}.

Magnetic Moment: is the couple exerted on a magnet placed at right angles to a uniform field with unit flux density. Unit, *ampere metre*^{2}.

Intensity of Magnetisation: is the magnetic moment per unit volume of a magnet. Unit, *ampere metre*$^{-1}$.

Pole Strength: unit magnetic pole is defined as one which when situated 1 metre distant in vacuum from an equal pole, experiences a force of $\mu_o/4\pi$ newtons. Alternatively, it can be defined as that pole strength which when placed in a unit induction experiences a force of 1 newton. Unit, *ampere-metre*.

Line of Force: A line of force is a curve in a magnetic field, such that the tangent at every point is the direction of the magnetic force at that point.

Magnetomotive Force: is defined as the line integral ∫H.dl evaluated for a closed path. It is equal to the total conduction current linked. Unit, *ampere*.

Coulomb's Magenetic Law: states that the force between two poles P_1 and P_2 situated distance *d* apart is given by $F=\mu_r\mu_o P_1P_2/4\pi d^2$,where μ_r is the permeability of the medium and μ_o the permeability of the free space = $4\pi x 10^{-7}$ henry metre^{-1}.
$\mu_r\mu_o$ replaces the permeability μ of the cgs system.

2.9.4 Thermal Units

The Degree Celsius (°C): The centigrade scale of temperature used the ice point as zero and the boiling point of water at 1 standard atmosphere as the upper fixed point set to be 100°C. The Celsius scale of temperature is defined to be the same as the thermodynamic scale with the zero shifted to the ice point, which is 273.15K, and thus:

θ/°C = T/K-273.15

The Degree Fahrenheit (°F): On the Fahrenheit scale, the ice point is 32°F and the steam point, 212°F. Thus t/°F = 32+1.8(θ/°C).

The Degree Reaumur (°R): On the Reaumur scale, the ice point is 0°R and the steam point, 80°R. Thus t/°R =0.8(θ/°C).

Quantity of Heat: Quantities of heat are measured in joules (J) in SI. Other units have been used, notably the *calorie*. The calorie is the amount of heat required to raise the temperature of 1 gramme of water by 1°C. This definition is not very precise, however, as the specific heat capacity of water varies with temperature, the 15° *calorie* was defined as the heat required to raise the temperature of 1 g of water from 14.5°C to 15.5°C. The *mean calorie* was defined as one hundredth of the heat required to raise the temperature of 1 g of water from 0°C to 100°C. The *kilocalorie* (1,000 calories) has also been used and written *calorie*. Where quantities of heat are expressed in calories, it is recommended that the conversion factor to convert to joules be stated.

In the FPS system, the *British thermal unit* (Btu) is used. This is the quantity of heat required to raise the temperature of 1 lb of water through 1^0F. The *therm* is 10^5 Btu.

Specific Heat Capacity (c_p,c_v): This is the amount of heat required to raise the temperature of 1 kg of a substance through 1K. Units, J kg^{-1} K^{-1}.

Molar Heat Capacity (C_m): This is the amount of heat required to raise the temperature of 1 mol of substance through 1K. Units, $Jmol^{-1}K^{-1}$.

Heat Capacity (C): The amount of heat required to raise the temperature of a body through 1K. Units, JK^{-1}.

Water Equivalent: The mass of water having the same total heat capacity as the given body.

Thermal Conductivity (λ): The rate of flow of heat (dQ/dt) through a surface of area, *A* in a medium is given by:

$$dQ/dt = -\lambda AdT/dx$$

where (dT/dx) is the temperature gradient, measured in the direction normal to the surface. The quantity λ, is the thermal conductivity of the medium. Units, $Jm^{-1}s^{-1}K^{-1}$, or $Wm^{-1}K^{-1}$.

Specific Latent Heat: The specific latent heat of fusion (specific enthalpy change on fusion) of a body is the heat required to convert 1 kg of the solid at its melting point into liquid at the same temperature. Unit, J kg^{-1}.

The specific latent heat of vaporization (enthalpy change on vaporization) of a liquid is the heat required to convert 1 kg of the liquid at its boiling point into vapour at the same temperature. Unit, J kg^{-1}.

Linear Expansivity: The increase in length per unit length per unit rise in temperature. Unit, K^{-1}.

Cubic Expansivity: The increase in volume per unit volume per unit rise in temperature. Unit, K^{-1}.

Critical Temperature (T_c): of a gas or vapour is that temperature above which it is not possible to liquefy the gas by the application of pressure alone. To liquefy a gas, it must be cooled below its critical temperature before being compressed.

Critical Pressure (p_c): That pressure which just liquefies a gas at its critical temperature.

Critical Volume (V_c): The volume of unit mass of gas at its critical temperature and pressure, i.e; it is the reciprocal of the critical density. It is often taken as the volume of one mole of a gas at its critical temperature and pressure.

Stefan-Boltzmann Law: The total energy, E of all wavelengths radiated per second per square metre by a full radiator at temperature T to surroundings at T_0 is given by $E = \sigma(T^4 - T_0^4)$, where σ is Stefan's constant. $\sigma = 5.669\,7 \times 10^{-8}$ W m^{-2} K^{-4}.

Planck's Radiation Law: The energy density of radiation in an enclosure at temperature T having wavelengths in the range λ to λ+dλ is $u_\lambda d\lambda$, where

$$u_\lambda d\lambda = 8\pi ch\lambda^{-5}(\exp hc/\lambda kT - 1)^{-1}\, d\lambda = c_1\lambda^{-5}(\exp c_2/\lambda T - 1)^{-1} d\lambda$$

$$c_1 = 4.992\,1 \times 10^{-24}\ \text{Jm} \quad ; \quad c_2 = 1.438\,79 \times 10^{-2}\ \text{mK}$$

The corresponding relation for radiation of frequency, v is

$$u_v dv = (8\pi h/c^3)(\exp hv/kT - 1)^{-1}\, v^3 dv.$$

where, h= Planck's constant; c = speed of light; k = Boltzmann's constant; T= temperature of the enclosure.

Wien's Displacement Law : The wavelength of the most strongly emitted radiation in the continuous spectrum from a full radiator is inversely proportional to the absolute temperature of that body, i.e; λT=b, where b is Wien's constant = 2.898×10^{-3} mK.

The Energy (E) of a quantum: of radiation of frequency v is $E = hv$ where h is Planck's constant.

2.9.5 Photometric and Optical Units

Luminous intensity: In SI, the unit of luminous intensity is the candela. This unit replaces the International candle which was defined in terms of the light emitted per second to all directions by a specified electric lamp (1 *candela* = 0.982 international candles). The values of *stilb* (*Sb*), *apostilb* (*asb*) and *lambert* (*L*) are frequently given not on the basis of the SI *candela*, but on the basis of the *international candle*, which is 1.005 times greater. Hence, the factors converting the *nit* into *Sb*, *L* should be divided by 1.005 (9.95×10^{-5} *Sb*, 3.13 *asb* and $3.13\times10^{-4}L$), and the factors *asb* and converting *stilb, apostilb* and *lambert* into *nit* multiplied by 1.005 (1.005×10^4 *nit*, 0.320 *nit* and 3.20×10^3 *nit*).

Einstein: is defined as unit of quantity of light equal to Avogadro's number of photons.

Luminous flux: The unit of luminous flux, the *lumen* (lm) is defined as the light energy emitted per second within unit solid angle by a uniform point source of unit luminous intensity. Thus 1 cd = 1 lm sr^{-1}.

Illuminance: of a surface is defined as the luminous flux reaching it perpendicularly per unit area. The British unit is the *lumen* ft^{-2}, formerly called the foot candle. The metric unit is the lumen m^{-2} or lux (lx).

Lambert's Cosine Law: For a surface receiving light obliquely, the illumination is proportional to the cosine of the angle which the light makes with the normal to the surface.

Brightness: of a surface is that property by which the surface appears to emit more or less light in the direction of view. This is a subjective quantity. The corresponding physical measurement of the light actually emitted is called the luminance.

Luminance: of a surface is the measure of the light actually emitted (i.e; the luminous intensity) per unit projected area of surface, the plane of projection being perpendicular to the direction of view. Unit, cd ft^{-2} or cd m^{-2}. In engineering, the luminance of an ideally diffusing surface emitting or reflecting one lumen ft^{-2} is called one *foot-lambert*.

The Refractive index: of a material is the ratio of the velocity of light in free space to that in the material.

Snell's law: For light incident on a boundary between two media, the ratio of the sine of the angle of incidence (the angle between the light ray in the first medium and the normal to the boundary surface) to the sine of the angle of refraction (the angle between the refracted ray in the second medium and the normal) is a constant, being equal to the inverse ratio of the refractive indices of the two media.

Dioptre: is the unit of measure of the power of a lens and is given numerically by the reciprocal of the focal length in metres.

2.9.6 Acoustical Units

Pressure : The unit of sound pressure is the *pascal* usually quoted as the root mean square pressure for a pure sinusoidal wave.

Frequency: The unit of frequency is the cycle per second, now designated as the *hertz* (Hz).

Threshold of Hearing: This is, for a normal (average) observer, the sound level or intensity which is just audible. For a pure sinusoidal note of frequency 1000 Hz, it is close to a root mean square pressure of 2×10^{-5} Pa.

Power Ratio: The unit of acoustical (or electrical) power measurement with respect to a standard level, is one *bel*. The interval between two powers W_l and W_0 in bels $\log_{10}(W_l/W_0)$. In practical work, the *decibel* (dB) is used. The interval between two powers W_l and W_0 is $10\log_{10}(W_l/W_0)$dB. In some instances, it is more convenient to employ natural logarithms. The power ratio so obtained is called the *neper*. The power interval between W_l and W_0 is $1/2\log_e(W_l/W_0)$ *nepers*. Hence 1 *neper* = 8.686 dB.

Intensity(I): Intensity of a sound wave in a given direction is the sound energy transmitted per second in that direction through unit area placed perpendicularly to the specified direction. Unit, Wm^{-2}. For a sinusoidal plane or spherical wave, the intensity is proportional to the mean square pressure exerted on an area at right angle to the given direction. Hence the interval between two intensities is given by $10\log_{10}(I_l/I_0)$dB or $20\log_{10}(p_l/p_0)$dB where p_l and p_0 are the root mean square pressures corresponding to the intensities I_l and I_0.

Loudness: Loudness is the physiological counterpart of acoustical intensity. It is a function of the intensity but also varies with frequency and composition of the note being heard. The *Weber-Fechner Law* states that the sensation (loudness) is proportional to the logarithm of the stimulus (intensity).

Loudness level: Loudness level of a sound is judged by comparison in free air with a standard sinusoidal note whose frequency is 1000 Hz. The unit is the *phon.* If an average observer decides that a sound is equally loud as the standard 1000 Hz note of intensity *n* dB above the standard reference level corresponding to a root mean square pressure of 2×10^{-5} Pa (i.e; the threshold of hearing), then the sound is said to have an 'equivalent loudness' of *n* British Standard *phons*.

Reverberation: Reverberation in an enclosure is the persistence of sound due to multiple reflections from the walls etc. of the enclosure.

Reverberation time: Reverberation time is the time required, from the moment of cessation of a sound for the intensity to drop by 60 dB, i.e; to one millionth of its original value. Unit, *second.*

Absorption Coefficient: Absorption Coefficient of a surface is the ratio of the sound energy absorbed to the total sound energy incident on the surface. The ideal absorber is one from which no sound is reflected or scattered. For unit area of various substances, the coefficient is expressed in terms of equivalent area of open window (diffraction effects excluded). Unit, *ft² of open window or Sabine*. The coefficient varies with frequency.

Sabine's Relation : For an auditorium whose walls etc. consist of areas S_1 S_2... etc. of absorption coefficient α_1 α_2 ... etc., the reverberation time *t* (in seconds) is given by $t = 0.05V/\Sigma\alpha S$ (unit of length, ft) or $t = 0.16V/\Sigma\alpha S$ (unit of length, metre) where V is the volume of the auditorium and $\Sigma\alpha S = \alpha_1 S_1 + \alpha_2 S_2 + ...$etc.

2.9.7 Radioactive Units

Radioactivity: The most common unit of radioactivity is the *Curie* (Ci). Originally defined as the volume of radon gas in equilibrium with 1 g radium, it has since become associated with the number of disintegrations occuring per second in 1 g of radium free from its daughter products, viz; 3.7×10^{10} disintegration per second. In modern usage, the curie has been redefined to agree with this result, and other units have been introduced as given below.

Curie: One *curie* (Ci) of any radioactive substance is that quantity in which 3.7×10^{10} atoms disintegrate per second. The millicurie (mCi) and microcurie (μCi) are in common usage.

The rutherford: Rutherford is the unit of activity corresponding to 10^6 disintegrations per second. Thus 37 rutherford = 1 mCi.

The roentgen: Roentgen was originally suggested as a unit of radiation and has become of universal use in defining the quantities of X-rays of γ-rays present. In 1937, the Fifth International Congress of Radiobiology recommended the following definition:

The *roentgen* is that quantity of X- or γ- radiation such that the associated corpuscular emission per 0.001293 g of dry air produces, in air, ions carrying 1 esu of quantity of electricity of either sign. This mass of air occupies 1 cm^3 at Standard Temperature and Pressure.

Dose rates: Radiation dose rate are often measured in units of roentgen $hour^{-1}$ or milliroentgen $hour^{-1}$ (mrh^{-1}).

The rad: Rad is defined as the absorbed dose of radiation when 1 g of material absorbs 100 ergs of energy. 1 rad = 10^{-2} J kg^{-1}.

The roentgen equivalent man (rem): It is the unit Dose Equivalent used in Radiation Protection. The Dose Equivalent is the product of the Absorbed Dose (measured in rad) and the quality factor Q, of the radiation. The value of Q indicates how damaging the particular radiation is, compared with 200 keV X- rays. Thus, low energy β-rays have Q = 1.7, while neutrons impinging on the eye have Q = 30.

A useful, but approximate formula for calculation of dose rates from γ-ray point sources is

$$\text{Dose rate (r hr}^{-1}) \simeq (5000\, C\, E)/d^2$$

where, C is the activity of the source in curies, E the energy of the γ-ray emitted in MeV and d is the distance from the source in cm. If more than one γ-ray is emitted, the total dose rate is the sum of the individual dose rates.

The law of radioactive decay: All radioactive substances transform at a rate which is proportional to the number of atoms present. If there are N_o atoms present at the zero of time, then at time, t, there are N_t atoms where,

$$N_t = N_o \exp - (\lambda t)$$

Here, λ, is a constant for the particular type of atom considered and is known as the transformation constant. The rate at which an atom decays is often measured in terms of the mean lifetime of the atom, τ, or the half-value period, $T_{½}$, which is often abbreviated to the half-life. The relation between these constants is :

$$\tau = 1/\lambda = T_{½}/\log_e 2$$

3

Units, Dimensions & Conversion Methods

3.1 Systems of Dimensions and Units

The systems of dimensions and units are built up by adopting a particular physical law as a basis . In mechanics, Newton's second law, that force is proportional to mass multiplied by acceleration, is chosen and the constant of proportionality is made unity so that

Force = Mass x Acceleration

and therefore,

Unit of force = Unit of mass x Unit of length/(Unit of time)^2

By specifying the units in which three of these quantities are to be measured, we can build up a complete, consistent system of units, applicable to any relationship derived mathematically on the assumption that the constant of proportionality in Newton's second law is unity.

Two main types of system are in use. The first, in which the fundamental units are *mass, length* and *time* so that the force is treated as a derived unit, is used in science. Such systems are called *absolute systems* and the fundamental units are defined by reference to particular objects or standards.

In the second type of system, the fundamental units are *force, length* and *time*. Such systems are known as *technical, engineering* or *practical systems*. For many purposes, engineers are concerned with forces and weights rather than masses and it was therefore, more natural to use the unit of force or weight as the fundamental unit.

Initially, such systems were gravitational systems in which the unit of force was defined as the force due to local gravity acting on the absolute unit of mass. Thus, in the British technical F.P.S. system, the unit of force is the weight of one pound of mass, and produces an acceleration of g ft/s^2 in 1 lb mass. Thus, the gravitational units of force are g times the absolute units of force. Since g is variable, such systems are not very convenient and gravitational units have been replaced by technical units of constant value based on an assumed standard value g_0 for gravitational acceleration of 9.80665 m/s^2 or 32.1740 ft/s^2. Thus, the weight of the body varies according to the local value of the actual gravitational acceleration g.

Unfortunately, several different units of mass, length, time and temperature are employed in different parts of the world and even in the same country for different purposes. This has produced a number of different systems of units some of which are consistent with a constant of unity in Newton's second law and others of which are not.

For Engineering calculations, at present, the most important system used is the International System of Units (Systéme International d' Unites; abbreviated in all languages as SI). Other systems of units employed by engineers, but now largely in the process of being replaced by SI Units, are the British foot-pound-second (F.P.S.) absolute system, the British F.P.S. technical system, the centimeter-gramme-second (C.G.S.) absolute and technical systems and the metre-kilogramme-second (M.K.S.) technical system. Different systems of measuring units are given in Table 3.1.

3.2 Conversion from One System to Another

Before inserting numerical values in any engineering formula, it is essential to ensure that these values are all in the same system of units.

Thus, for any calculation in British technical units, the numerical values of force must be in pound-force, since mass is in slugs, length

Table 3.1 : Different Systems of Measuring Units

Measurement	*International system (L,M,T) or MKS Mass system or SI*	*Absolute (L,M,T) or CGS system*	*Absolute (L,M,T) or British or FPS system*	*Gravitational (L,F,T) or British or FPS system*	*Gravitational (L,F,T) or MKS Force system*
Mass	kilogram (kg)	gram mass (g_m)	pound mass (lb_m)	slug (slug) (lb_f-sec^2-ft^{-1})	kg_f-sec^2-m^{-1}
Length	metre (m)	centimetre (cm)	feet (ft)	feet (ft)	metre (m)
Time	second (s)	second (sec)	second (sec)	second (sec)	second (sec)
Force	newton (N) (kg_m-m-sec^{-2})	dyne (dyn) (g_m-cm-sec^{-2})	poundal (Pdl) (lb_m-ft-sec^{-2})	pound force (lb_f)	kilogram force (kg_f)
Work	Joule (J)	erg (erg) (cm dyn)	ft Pdl	ft lb_f	m kg_f
Torque or moment of force	N m	dyn cm	Pdl ft	lb_f ft	kg_f m
Power	watt (W) (J s^{-1})	erg sec^{-1}	ft Pdl sec^{-1}	horse power (hp) lhp=550 ft.lb_f sec^{-1}	m kg_f sec^{-1} 1 metric hp (cheval-vapeur) = 75 mkg_f sec^{-1}
Remarks	Systéme International (SI) to be used universally.	Used by Physicists everywhere.	Frequently used in British technical writings and by physicists occasionally.	Used by Engineers in English speaking countries.	Used by Engineers in non-English speaking countries, continental Europe.

in feet and time in seconds. For example, pressure intensities must be expressed in pound-force per square foot.

To bring all these into the same system of units, the given numerical values have to be multiplied by appropriate conversion factors. This presents no difficulty with the fundamental units and the simpler derived units, but for the more complex quantities, a knowledge of the dimensions of the quantity can be of value in calculating the required conversion factor. The true physical value of any quantity must be independent of the system of units employed to measure it. If a quantity Q is found to have a numerical value of n_1 when measured in units of size u_1,

Physical value of $Q = n_1u_1$

Similarly, if Q is found to have a numerical value of n_2 when measured in units of size u_2,

Physical value of $Q = n_2u_2$

Since the value of Q must be the same in both cases;

$Q = n_1u_1 = n_2u_2$

Thus, in general, the numerical value of a quantity is inversely proportional to the size of the unit in which it is measured.

If the quantity Q has a dimensional formula $[M^aL^bT^c]$, the unit of measurement will be a derived unit (M, L and T are dimensional notations for mass, length and time, respectively). Suppose that the fundamental unit of mass, length and time are m_1, l_1 and t_1 in the first system and m_2, l_2 and t_2 in the second system, and that the size of the unit m_1 is k_Mm_2, of l_1 is k_Ll_2 and of t_1 is k_Tt_2. The derived units based on the dimensional formula will be $u_1 = (m_1^a)(l_1^b)(t_1^c)$ and $u_2 = (m_2^a)(l_2^b)(t_2^c)$.

Suppose that the numerical value of Q is N_1 when measured in the first system & N_2 when measured in the second system, then

Physical value of $Q = N_1u_1 = N_2u_2$

Now $N_1u_1 = (N_1)(m_1^a)(l_1^b)(t_1^c)$ therefore, replacing m_1 by k_Mm_2, l_1 by k_Ll_2 and t_1 by k_Tt_2

$N_1u_1 = N_1(k_M^a m_2^a)(k_L^b l_2^b)(k_T^c t_2^c)$

$$= N_1 k_M^a k_L^b k_T^c \times m_2^a l_2^b t_2^c$$

But $u_2 = m_2^a\, l_2^b\, t_2^c$, therefore;

$$N_1 u_1 = (N_1\, k_M^a\, k_L^b\, k_T^c)\, u_2 = N_2 u_2$$

So that $N_2 = N_1\, k_M^a\, k_L^b\, k_T^c$

where k is the number of the units in the second system required to make one of the fundamental units in the first system.

Example: 3.2.1

The mass density of water is 62.4 lb/ft³ in F.P.S. absolute units; what are the corresponding values in (a) C.G.S. absolute units, and (b) SI Units.

The dimensional formula for mass density is $[ML^{-3}]$, using suffix 1 for F.P.S. absolute units and suffix 2 for the alternative units, the numerical value of the mass density N_2 in the alternative units corresponding to a value N_1 in British absolute units is given by;

$$N_2 = N_1\, k_M\, k_L^{-3}$$

(a) For mass, the unit in the first system is the pound-mass and in the second, the gramme-mass, so that k_M is the number of gramme-mass required to make 1 pound-mass & is therefore 453.6.

For length, the unit in the first system is the foot and in the second the centimetre so that k_L is the number of centimetres in 1 ft and is therefore 30.48. Therefore,

$$N_2 = N_1 \times 453.6 \times 30.48^{-3}$$

$$= 62.4 \times 453.6/(30.48)^3 = 1.0 \text{ g/cm}^3$$

i.e; the mass density of water is 1 g/cm³ in the C.G.S. absolute system, and the conversion factor N_1/N_2 is 62.4

(b) For mass, the unit in the first system is the pound mass and in the second system, the kilogramme mass, so that k_M is the number of kilogramme mass required to make one pound mass, which is 0.4536.

For length, the unit in the first system is the foot and in the second the metre, so that k_L is the number of metres corresponding to one foot, which is 0.3048. Therefore,

$$N_2 = N_1 \times 0.4536 \times 0.3048^{-3}$$

$$= 62.4 \times 0.4536 / 0.3048^3 = 1000 \text{ kg/m}^3$$

i.e., in the SI, the mass density of water is 1,000 kg/m³.

Example: 3.2.2

What is the conversion factor required to change the value of thermal conductivity from British thermal units per foot, hour and degree Fahrenheit into SI units (joules per metre, second and degree celsius).

The units employed for the quantities in the question correspond to the HLTθ (H and θ are notations for heat quantity and temperature, respectively) system of dimensions, which is therefore the obvious system to use in the determination of the conversion factor.

In the HLTθ system, the dimensions of thermal conductivity are $[HL^{-1}\ T^{-1}\ \theta^{-1}]$

Using suffix 1 for British units and suffix 2 for SI units

$$N_2 = (N_1)(k_H)(k_L^{-1})(k_T^{-1})(k_\theta^{-1})$$

where, k_H, k_L, k_T, k_θ are the ratios of the units of heat, length, time and temperature in system 1 to the corresponding units in system 2, as tabulated below :

Quantity	*System (1), British*	*System (2), SI*	*Ratio (1)/(2)*
Heat	Btu	joule	1,055
Length	foot	metre	0.3048
Time	hour	second	3,600
Temperature	deg F	deg C	0.556

therefore, $N_2 = N_1 \times 1{,}055 / (0.3048 \times 3600 \times 0.556) = 1.73\ N_1$,

value in SI Units = 1.73 × value in Btu/ft.h.deg F

Example: 3.2.3

An engine produces 57 hp. What is the corresponding value in kilowatts and what is the conversion factor?

The horsepower and the kilowatt are multiples of the basic units for power in the British technical system and the SI, respectively.

Thus, if n_1 is the horsepower and N_1 is the corresponding number of British technical units (ft lb_f/sec)

$$N_1 = 550\ n_1$$

Similarly, since the basic unit of power in the SI is the watt (W),

$$N_2 = 1{,}000\ n_2$$

The dimensional formula for power is $[ML^2T^{-3}]$

$$N_2 = (N_1)(k_M)(k_L^{\ 2})(k_T^{\ -3})$$

where, k_M, k_L and k_T are the ratio of the units of mass, length and time in system 1 to the corresponding units in system 2, as tabulated below:

Quantity	*System (1),FPS technical*	*System (2), SI*	*Ratio (1)/(2)*
Mass	slug	kilogramme	14.6
Length	foot	metre	0.3048
Time	second	second	1.0

Therefore, $N_2 = N_1 \times 14.6 \times (0.3048)^2 = 1.356\ N_1$

or $1{,}000\ n_2 = 550\ n_1 \times 1.356$

so that $n_2 = 0.746\ n_1$

Value in kilowatts = 0.746 × value in horsepower

Putting n_1 = 57 hp, output = 0.746 × 57 = 42.4 kW

Conversion factor; $n_2/n_1 = 0.746$

Example: 3.2.4

The force required for cutting in a lathe is given by F=KA, *where F is the tangential force (lb_f), K is a constant for the material used, and A is the area of the cut (in^2). Derive an expression for a constant K_{SI} in SI units in terms of K_{fps} in F.P.S. units so that if A is expressed in mm^2 then F will be given in newton units.*

Rearranging the given equation to obtain the units of K_{fps}

$$K_{fps} = F/A\ (lb_f/in^2)$$

$$K_{SI} = K_{fps}\ (lb_f/in^2) \times [4.448N/lb_f] \times [in^2/25.4^2\ mm^2]$$

$$= 4.448/25.4^2 \times K_{fps}\ (N/mm^2)$$

$$\text{i.e.; } K_{SI} = 0.0069\ K_{fps}\ N/mm^2$$

Example: 3.2.5

Dynamic viscosity (η) is given by (shear stress)/(velocity gradient), expressed in units of $lb_f s/ft^2$ or Ns/m^2. Derive an expression for η_{SI} in terms of SI units, given values of η_{fps} in F.P.S. units.

$$\eta_{SI} = (lb_f s/ft^2) \times [4.448N/lb_f] \times [ft^2/0.3048^2 m^2]$$

$$\eta_{SI} = 47.9\ \eta_{fps}\ (Ns/m^2)$$

3.3 Conversion of Dimensional Constants

Many equations used in practice contain numerical constants which are not dimensionless. Such equations will only be dimensionally homogeneous if the correct dimensions are assigned to the constants, and the values of the constants will vary according to the system of units employed.

Example: 3.3.1

The velocity, v, of flow in an open channel can be calculated from the Chezy formula, $v=c\sqrt{(mi)}$, where C is a constant with a numerical value depending on the surface finish of the sides and base of the channel; m is the hydraulic mean depth, defined as the ratio of the cross-sectional area of flow, A_l to the length, P_l of the wetted perimeter of the cross-section; and i is the slope of the channel, i.e; the drop per unit length. What are the dimensions of

C? If C has a value of 120 in the F.P.S. technical system what is the corresponding value in the M.K.S. absolute system (SI)?

Since $v = C\sqrt{(mi)}$,

$$C = v/\sqrt{(mi)} = vp^{1/2}/[(A)^{1/2}(i)^{1/2}]$$

The dimensions of the quantities on the right hand side of this equation are;

velocity $[v] = [LT^{-1}]$

Perimeter $[p] = [L]$

Area $[A] = [L^2]$

slope $[i] = [L\,L^{-1}] = 1$

For dimensional homogeneity, the dimensions of C must be the same as those of the right hand side:

Dimensions of $C = [LT^{-1}]\,[L]^{1/2}/[L^2]^{1/2} = [L^{1/2}T^{-1}]$

Using suffix 1 for the F.P.S. system and suffix 2 for the MKS system;

$$N_2 = N_1\,(k_L)^{1/2}(k_T)^{-1}$$

where, k_L and k_T are the ratios of the units of length and time, respectively in system 1 to the corresponding units in system 2. The units are

Quantity	*System (1),FPS technical*	*System (2), SI*	*Ratio (1)/(2)*
Length	foot	metre	0.3048
Time	second	second	1.0

Therefore, $N_2 = N_1 \times 0.3048^{1/2} \times 1$

Putting $N_1 = 120$

$N_2 = 120 \times 0.3048^{1/2} = 66.2\text{m}^{1/2}/\text{sec}$

This is the required value of C.

3.4 Use of Dimensional Reasoning to Check Equations

The principle of dimensional homogeneity requires that any equation which fully and correctly represents a possible physical situation must be dimensionally homogeneous so that each term has the same dimensional formula. It is, therefore, possible to make a rapid check of any algebraic analysis of such a situation by substituting the dimensions of the quantities in each term and checking that in the resultant dimensional equation each fundamental dimension appears to the same power in each term. This is only a partial check since dimensional analysis will give no indication as to whether the equation is numerically correct, but it will show whether the algebraic manipulation of the quantities has been carried out correctly and so prevent incorrect formulation of equations. Many engineering formulae in day to day use are committed to memory.

Example: 3.4.1

The quantity of fluid flow Q (volume per unit time) is given by the equation Q=A.v where, A is the area of cross-section and v is the velocity of fluid flow. Verify the equation for dimensional homogeneity ?

The given equation is : Q=A.v

Dimensions of L.H.S. are :

$$Q \text{ (quantity of fluid flow)} = L^3/T = L^3\ T^{-1}$$

Dimensions of R.H.S. are :

$$A.v \text{ (area x velocity)} = L^2.L/T = L^3\ T^{-1}$$

Hence, the L.H.S. and R.H.S. are equal dimensionally. Any other combination of the variables Q,A and v will not be dimensionally homogeneous. Thus, dimensional homogeneity can be used to verify physical equations.

Example: 3.4.2

The thermal diffusivity (cm²/s), α of a material is given by the equation α=k/(ρ.c) where k is the thermal conductivity (cal/cm.s.°C) of the material, ρ is it's bulk density (g/cm³) and c is the specific heat (cal/g.°C) of the material. Show the validity of the equation using dimensional reasoning ?

The given equation is : $\alpha = k/\rho.c$

Dimensions of L.H.S. are :

α (thermal diffusivity)= α (cm^2/s) = $L^2/T = L^2 T^{-1}$

Dimensions of R.H.S. are :

$k/\rho.c$ {thermal conductivity/(bulk density × specific heat)}

$= k/\rho.c\ [(cal/cm.s.^0C)/\{(g/cm^3) \times (cal/g.^0C)\}]$

$= [\{(ML^2/T^2)/L.T.\theta\}/\{(M/L^3) \times ((ML^2/T^2)/M.\theta)\}]$

(where ML^2/T^2 are the dimensions of cal, since, 1 cal=4.19 joule=4.19 Newton.metre =(kg.m/s2).m=ML^2/T^2)

$= [\{ML^2T^{-2}.L^{-1}T^{-1}\theta^{-1}\}/\{ML^{-3}ML^2T^{-2}M^{-1}\theta^{-1}\}]$

$= [\{MLT^{-3}\theta^{-1}\}/\{ML^{-1}T^{-2}\theta^{-1}\}]$

$= L^2T^{-1}$

Hence, the L.H.S. and R.H.S. are equal dimensionally. Any other combination of the variables α, k, ρ and c will not be dimensionally homogeneous. Thus, the dimensional reasoning can be used to show the validity of a physical equation.

4

Some Selected Units of Measurements Which are/were Used in Different Countries and their Equivalents

Unit	*Country/Place*	*Equivalent*
Almude	Portugal	16.7 litres
	Spain	4.625 litres=1/12 fanega (dry measure)
Angstrom unit (spectroscopy)		10^{-10} metre=1/10,000 micron
Anker	Latvian S.S.R.	38.256 litres, or 30 stoof
Anoman (ammomam, amomam)	Ceylon (Sri Lanka)	5.77 U.S. bushels
Archin or arshin	Turkey	1 new archin (Law 1881) = 1 metre (39.37 inches) = 10 parmaks (decimetres) = 100 khats (centimeter), 1 mill= 1,000 archins (kilometre), pharoagh = 10 mills; another pharoagh = 2 hours journey
Ardeb	U.A.R.	5.619 bushels
Are		100 square metres=119.6 square yards

Unit	Country/Place	Equivalent
Arpent	France	legal arpent was equal to 100 square perches=51.07 metric ares;
	Canada (Quebec)	180 French feet
Arroba	Portugal	14.68 to 15 kilograms
	Spain	wine=16.14 litres; oil=12.56 litres
Arshin	U.S.S.R. Estonian S.S.R.	28 inches, or 0.7112 metre
Artaba	Iran	66.0 litres
Assayton	U.S.	29.167 grams
Aune	Belgium	1 metre; formerly 1.2 metres
	France	1.188 metres
	Jersey	4 feet
Barile	Rome	58.34 litres
Bat, Baht or tical	Thailand	15 grams (=231.5 grains)
Batman	Iran	6½ pounds avdp; varies locally
	Turkistan	125 kilograms (variable)
Behär or Bahar	Arabian Peninsula	450 pounds avdp.
Berri	Turkey	1.67 kilometres (1.04 miles) (old measure)
Boisseau	Belgium	15 litres
Boutylka	U.S.S.R.	1.625 U.S. liquid pints
Braca	Portugal	2.20 metres
Braccio d'ara	Rome	0.7 metre
Brasse	France	1.62 metres
Braza	Argentina	1.732 metres
Bu or tsudo	Japan	100/30.25=3.306 square metres (area); also = 1/100 shaku (length)

Unit	*Country/Place*	*Equivalent*
Bunder	The Netherlands	1 hectare
Cabot	Jersey	10 pots, or 19.75 litres
Candy	Bombay (now Mumbai)	560 pounds avdp.
	Madras (now Tamil Nadu)	500 pounds avdp.
Cantor or kantar	Turkey	124.45 pounds avdp. (old weight)
Cantara	Spain	1 arroba
Capicha	Iran	263 litres
Carat, metric		200 milligrams
Catty	China	$1^1/_3$ pounds avdp. (see Tael)
	Malaysia	$1^1/_3$ pounds avdp.
	Thailand	2.67 pounds avdp. (old weight)
Cawnie	Madras (now Tamil Nadu)	1.322 acres
Centner	Austria, Denmark, Switzerland	50 kilograms
Chain	Great Britain	66 feet
Chain(Gunter's)	U.S.	66 feet
Chang	China	10 ch'ih=11 feet 9 inches*
	Thailand	1,200 grams
Chapah	North Borneo (now Sabah)	1.8 pounds avdp.
Charka chee (see Tahil)	U.S.S.R.	0.123 litre

* *values agreed on by China in treaties (1842-44 and 1858-60) with England and France and therafter accepted by the Chinese maritime customs as standard values for all tariff duties but not in general otherwise used.*

Unit	Country/Place	Equivalent
Chek	Hong Kong	$14^5/_8$ inches
Chenica	Iran	1.359 litres
Chetvert	U.S.S.R.	8 chetveriks, or 2.099 hectolitres
Ch'ien	China	$58^1/_3$ grains (silver weight)
Ch'ih	China	equivalent varies throughout China from 11 to 15.8 inches; for customs purposes the Treaty ch'ih=14.1 inches, and 5 ch'ih = 1 pu
Chin or Catty	China	$1^1/_3$ pounds avdp.
Ching	China	121 square feet
Ch'ing	China	72,600 square feet
Chittack or chhattack	Bengal	5 tolas, or 900 grains
Chô	Japan	as unit length=360 shaku as unit area=3,000 bu or tsubo
Chiio	China	1,815 square feet
Chupah	Singapore	3.125 pounds avdp of water at 62^0 F, as a measure of capacity
	Malacca	36 ounces avdp of water
Chupak	Malaysia	1 quart (British Imperial)
Colluthun	Iran	8.22 litres
Coss	India	1.829 kilometres
Côvado	Portugal	0.66 metre
Covid or cubit	Madras (now Tamil Nadu)	18 to 21 inches
	Bombay (now Mumbai)	18 inches
Covido	Arabian Peninsula	19 inches

Unit	*Country/Place*	*Equivalent*
Cuartillo	Spain	1.16 litres (dry); 0.504 litre (liquid)
Daktylos	Greece	25.4 millimetres
Daribah	U.A.R.	15.84 hectolitres
Deciatina	U.S.S.R.	2,400 square sagenes=2.7 acres
Denaro	Italy	1 gram
	Rome	18.17 grains (old weight)
Deunam	Turkey	25 ares
Diraa or pik	U.A.R.	0.58 metre
	Turkey	0.7087 metre
Dirhem	U.A.R.	1.761 drams avdp. (customs) 3.12 grams (Cairo)
Dito	Italy	1 centimetre
Djerib	Turkey	1 hectare
Dolia or dola	U.S.S.R.	0.686 grain 96 doli=1 zolotnick
Drachm	Great Britain	3 scruples (apothecaries' weight)
Drachm, fluid	Great Britain	0.961 U.S. fluid dram
Drachma	The Netherlands	3.906 grams (apothecaries' weight)
	Turkey	3.21 grams
	Greece	3.2 grams
Dram	Great Britain	1/16 avoirdupois ounce
	Iran	1 gram
Ducat	Vienna	53.873 grains (gold weight)
Duim	The Netherlands	1 inch
	U.S.S.R.	1 inch
Eimer	Germany	29 to 307 litres

Unit	*Country/Place*	*Equivalent*
El	The Netherlands	1 metre(old el=27.08 inches)
Ell	Jersey	4 feet
Ella	North Borneo (Sabah)	1 yard
Elle	Latvian S.S.R.	0.537 metre
	Switzerland	60 centimetres
Estadio	Portugal (old)	258 metres
Faden	Latvian S.S.R.	4.077 steres
Faltche	Moldavian S.S.R.	143.22 ares
Fanega	Argentina	137 litres
	Portugal	55.364 litres
	Spain	50.5 litres
	Peru	50.5 litres
Feddan	U.A.R.	1.038 acres (Masri)
Fen	China	5.76 grains (silver weight)
Fjerding	Denmark	0.988 bushel
Fod	Denmark	0.3138 metre = 1.0297 feet
Fot	Sweden	11.689 inches; 10 fot=1 stöng; 1 ref= 10 stänger; 1 mil=360 ref
	Norway	0.3137 metre
Founte or funt	U.S.S.R.	0.90282 pound avdp
Foute or fut	U.S.S.R.	1 English foot
Frasco	Argentina	$2^3/_8$ litres
Funt	Poland	405.504 grams
Furlong	U.S.;Great Britain	220 yards
Fusz	Vienna	12 zolls=1.037 feet
	Switzerland	$1^1/_3$ fusz=1 metre (see Stab)
Gantang	Malaysia (West)	1 British Imperial gallon

Unit	*Country/Place*	*Equivalent*
	North Borneo (Sabah)	144 ounces avdp. weight of water as measure of capacity
Garnetz	U.S.S.R.	3.28 litres
Gin (*see Kati*)		
Gisla	Zanzibar	measure of 360 pounds avdp of rice
Gô	Japan	0.18 litre
Grain	U.S.;Great Britain	1/7000 avoirdupois pound =0.065 gram
	U.S.S.R.	0.96 grain (apothecaries')
Grano	Rome	0.757 grain
Grao	Portugal	0.769 grain; also measure 0.18 inch
Grein	The Netherland	0.065 gram
Guz or gudge	India: Bengal	36 inches
	Mumbai	27 inches
	Madras (now Tamil Nadu)	33 inches
	Iran	the guz, gueza, or zer varies from 24 to 44 inches, a guz of 40.95 inches is common; by the law of 1925, the guz was made equal to the metre
	Arabian Peninsula	25 inches
Hat'h or moolum or cubit	India	18 inches
Hiyaka-me	Japan	5,797,198 grains
Hiyak-kin	Japan	132 ½ pounds avdp.
Hoon (*see Tahil*)		
Hu	China	51.77 litres
Immi	Switzerland	1.5 litres

Unit	*Country/Place*	*Equivalent*
Jaan	India (Tamil Nadu)	Measure from tip of middle finger to tip of thumb, equivalent to 9 inches.
Joch	Austria	57.55 ares=1.422 acres
Kan	The Netherlands	1 litre
	Hong Kong	$1\frac{1}{3}$ pounds avdp.
Kanne or kanna	Germany	1 litre, or formerly 1.06 U.S. liquid quarts, or 0.91 U.S. dry quart
	Sweden	2.62 litres
Kantar or cantaro	U.A.R.	99.0492 pounds avdp=100 rotls (customs); 45 kilograms of cotton; 44.5 kg other produce
Karwar	Iran	100 batmans
Kassabah	U.A.R.	3.8824 yards
Kati, catty, or gin	Malaysia	$1\frac{1}{3}$ pounds avdp.
Keddah	U.A.R.	2.0625 litres
Ken	Japan	1.82 metres=5.965 feet
	Thailand	40 inches
Kerât	Turkey	$1\frac{1}{8}$ inches measure (old) 3.09 grains weight (old)
Kette or chain	Germany	10 metres
Khat	Turkey	1 centimetre
Kileh	Turkey	1 bushel approximately (varies)
Kin	China	see Catty
	Japan	0.600 kilogram=1.323 pounds
Klafter	Austria	2.074 yards
	Switzerland	1.9685 yards
Köddi	Arabian Peninsula	7.58 litres
Koilon	Greece	33.17 litres

Unit	*Country/Place*	*Equivalent*
Koku	Japan	180.39 litres (see *To*)
Kon or catty	Korea	$1\frac{1}{3}$ pounds avdp.
Kortonde	Norway	138.97 litres
Korn-topmaal	Norway	160 litres
Korrel	The Netherlands	0.1 gram
Kouza	Cyprus	9 British Imperial quarts
Koyan	Malaysia	$5{,}333\frac{1}{3}$ pounds avdp.
Krina	Bulgaria	20 litres
Kulmet	Latvian S.S.R.	10.93 litres
Külimet	Estonian S.S.R.	11.48 litres
Kung	China	78.96 inches
Kup	Thailand	10 inches
Kwan or kuwan	Japan	3.75 kilograms=8.267 pounds avdp.
Kwarta	Poland	1 litre
Kyat	Burma	16.33 grams
Last	The Netherland	30 hectolitres
Latro	Czechoslovakia	1.917 metres
Li	China	about $\frac{1}{3}$ mile=360 pu; varies with length of ch'ih 1 decigram
Liang or léang	China	$1\frac{1}{3}$ ounces; 16 liang=1 chin $=1\frac{1}{3}$ pounds avdp.
Libbra	Italy	1 kilogram; 339 grams (old)
Libra	Argentina	1.0128 pounds avdp.
Libra (Castilian)	Spain, Mexico	1.0143 pounds
Libra or arratel	Portugal	1.012 pounds avdp.
Light year (astronomy)		9.4614×10^{12} kilometres
Line or ligne	Paris	2.26 millimetres

Unit	*Country/Place*	*Equivalent*
Liniya	U.S.S.R.	0.1 inch; 1 archine=280 liniyas
Link, Gunter's or surveyors'	U.S. Great Britain	7.92 inches
Litre		0.908 U.S. dry quart, or 1.057 U.S. liquid quarts
Livre	Belgium	1 kilogram
	France	0.4895 kilogram
Loket	Czechoslovakia	0.593 metre (Prague); 0.594 metre (Moravia); 0.579 metre (Silesia)
Lot	Germany	new lot=10 grams; old lot, nearly ½ ounce avdp.
	Switzerland	15.625 grams
	Vienna	270.1 grains; postal lot 257.2 grains
Maass	Germany	1.837 litres
	Switzerland	1.5 litres
Maatje	The Netherlands	0.1 litre
Mace	China	$58^1/_3$ grains
	North Borneo (Sabah)	$93^1/_3$ grains
Mahnd	Arabian Peninsula	2.04 pounds avdp.
Marc or mark	France	0.2448 kilogram (old weight)
	Sweden	6.77 ounces troy
	Vienna	9.02 ounces troy
Marco	Portugal	8 oncas=229.5 grams
	Spain	3,550 grains
Maund	India	82.286 pounds avdp, government; 28 pounds, Bombay (Mumbai); 25 pounds,

Unit	*Country/Place*	*Equivalent*
		Madras (Tamil Nadu); local maunds vary
Metze	Austria	61.5 litres
Miglio	Rome	0.925 mile
Mijl	The Netherlands	1 international nautical mile
Mil	U.S.;Great Britain	0.001 inch
Mil or mill	Turkey	1,000 archins (new mil)
	Denmark	4.680 miles
Mile, nautical	U.S.	6,080.20 feet (formerly)
Mile, nautical ("admiralty")	Great Britain	6,080 feet
Mile, nautical, international		1,853 metres
Mile (postal)	Austria	4.714 miles
Milha	Portugal	1.297 miles
Miscal	Iran	10 grams, formerly 71 grains
Mkono	East Africa	45.72 centimetres
Mna	Greece	1½ kilograms = 1.172 oke
Momme	Japan	1/1,000 kwan = 10 fun, 3.75 grams
Morgen	Prussia	0.631 acre
	The Netherlands (old)	2.1 acres
	South Africa	2.117 acres
Mou	China	commonly 806.65 square yards; varies locally; Shanghai=6,600 square feet (municipal council); by customs treaty=920.417 square yards, based on ch'ih of 14.1 inches.
Mud	The Netherlands	1 hectolitre

Unit	*Country/Place*	*Equivalent*
Mulam	India (Tamil Nadu)	2xJaan=18 inches=1 cubit.
Nin	Thailand	10/12 inch
Ock	Turkey	legal ock (1881)=100 drachmas; new batman=10 ocks, and kantar=10 batmans; ock=1 kilogram
Octavillo	Spain	0.29 litre
Oitavo	Portugal	1.730 litres
Oke	Bulgaria	1.28 litres (for liquids); 1.282 kilograms (old)
	Cyprus	2.8 pounds avdp=400 drams (Cyprus)
	U.A.R.	2.751 pounds avdp; 2.805 pounds (Alexandria)
	Greece	2.82 pounds=1.280 kilograms
	Turkey	1.28 pints; 2.83 pounds avdp (old)
Onca	Portugal	28.688 grams
Once	France	30.59 grams (old)
Oncia	Rome	436.165 grains
Ons	The Netherlands	100 grams
Packen	U.S.S.R.	1,083.38 pounds avdp.
Palamé	Greece	25.40 millimetres
Palm	The Netherlands	1 decimetre
Palmo	Portugal	0.22 metre
	Spain	0.209 metre
Para	North Borneo (Sabah)	90 pounds avdp.
Parah	Ceylon (Sri Lanka)	25.4 litres

Unit	*Country/Place*	*Equivalent*
Parasang		(*see Persakh*)
Parmak		(*see Archin*)
Parsec (astronomy)		3.084 x 10^{13} kilometres
Passeree or paseri	Bengal	5 seers
Pé	Portugal	$^1/_3$ metre
Pecheus	Greece	0.648 metre
Perch	U.S.	1 rod
	Great Britain	30¼ square yards
Perche	France	22 square pieds de roi;
	Canada	in Quebec 18 French feet
Persakh	Iran	6,000 guz=3.88 miles
Pfund	Germany	16 unzen=32 lot, 1.01 to 1.23 pounds avdp, old weight; zollpfund (1872)=500 grams
	Latvian S.S.R.	419 grams
	Switzerland	500 grams=16 unze apothecaries' pfund=375 grams
	Vienna	pfund=560 grams; zollpfund (1871)=500 grams
Pharoagh		(*see Archin*)
Pic or pik	Cyprus	2 feet
Picki, pik, or pic	Greece	0.64 to 0.67 metre
Picul	Japan	132.3 pounds avdp.
	Malaysia (West), Hong Kong	$133^1/_3$ pounds avdp.
	North Borneo (Sabah)	a measure of 180 pounds weight of water
Pié	Rome	11.73 inches
	Spain	10.97 inches
Pied	Belgium	11.81 inches= 10 pouces
	Canada	12.79 inches

Unit	*Country/Place*	*Equivalent*
Pied de roi	Paris	0.3248 metre
Pik		(*see Diraa*)
Pinte	France	0.931 litre
Pipa	Portugal	500 litres; 429 litres (old value)
Poide de marc	France	0.2448 kilogram=8 ounces
Polegada	Portugal	27.77 millimetres
Pond	The Netherlands	½ kilogram
Pood	U.S.S.R.	36.113 pounds
Pot	Denmark	0.966 litre
	Switzerland	1½ litres
	Belgium	1½ litres (dry); ½ litre (liquid)
	Norway	0.965 litre
Pounce	France	27.07 millimetres=1.066 inches (old measure)
Pu	China	70.5 inches=5 ch'ih
Puddee or padi	Madras (now Tamil Nadu)	1.64 litres; 100 cubic inches= government puddee
Pulgada	Spain	0.914 inch
Pund	Denmark	500 grams
	Norway	500 grams; 0.4981 kilogram (old value)
	Sweden	425.1 grams
Quarter	Great Britain	28 pounds; 8 British Imperial bushels=8.257 U.S. bushels
Quarto	Rome	73.6 litres
	Portugal	3.46 litres
Quintal	Spain	100 libras (Castilian)=101.4 pounds
	Portugal	58.752 kilograms, 129½ pounds avdp.
	Argentina	100 libras, or 101.28 pounds avdp.

Unit	*Country/Place*	*Equivalent*
Quintal (metric)		100 kilograins=220.46 pounds avdp.
Ratel	Iran	1.014 pounds avdp.
Rattel or rottle	Arabian Peninsula	1.02 pounds avdp.
Ri	Japan	2.440 miles
Rode	Denmark	3.138 metres
Roede	The Netherlands	1 dekametre
Rood	Great Britain	40 perches=¼ acre
Rotl, rottelo or ratel	U.A.R.	0.9905 pound avdp. (customs)
Rottol	Turkey	1.6 litres (old measure)
Sagene	U.S.S.R.	7 feet
Scheffel	Germany	50 litres, formerly 16 metzen (Prussia)
Schepel	The Netherlands	1 decalitre
Schoppen	Germany	½ litre
	Switzerland	0.375 litre
Sê	Japan	118.615 square yards (0.9918 are)
Seer*	India	government seer=2.057 pounds avdp; Bengal, 80 tolas weight of rice (heaped measure), about 60 cubic inches (struck measure);
		Madras (now Tamil Nadu), 0.617 pound avdp;
		Bombay (now Mumbai), 0.72 pound avdp.
	Ceylon (Sri Lanka)	1.06 litres

* *In India the seer , like the maund, varies considerably: usually 40 seers equal a maund.*

Unit	*Country/Place*	*Equivalent*
	Iran	16 miscals, or 1,136 grains weight (Sihr)
Seidel	Austria	0.354 litre
Sen	Thailand	40.6 metres
Shaku	Japan	10/33 metre, also 3.306 square decimetres; also 18.039 cubic centimetres (1/100 shô)
Sheng	China	1.035 litres
Shih	China	157.89 pounds
Shô	Japan	1.804 litres
Skaal-pound	Sweden	425.1 grams, or 0.937 pound avdp.
	Norway	0.4981 kilogram, or officially ½ kilogram
Skeppe	Denmark	17.39 litres
Skjeppe	Norway	17.37 litres
Stab	Germany	1 metre, or $3^1/_3$ old fuss, but varied
Stere		1 cubic metre
Stone	Great Britain	14 pounds avdp.
Stopa	Poland	0.288 metre
Streep	The Netherlands	1 millimetre
Stremma	Greece	10 ares
Strich	Germany	1 millimetre
	Switzerland	$3^1/_3$ strich= 1 millimetre
Stunde	Switzerland	4.8 kilometres
Sultchek	Turkey	cubic measure (1881) whose sides equal a parmak (decimetre)
Tael	Thailand	936¼ grains
	Hong Kong	$1^1/_3$ ounces avdp.

Unit	*Country/Place*	*Equivalent*
	China	silver weight, $1^1/_3$ ounces avdp.
	Japan	10 momme
Tahil	Malaysia	$1^1/_3$ ounces avdp=10 chee= 100 hoon
Tan	China	$133^1/_3$ pounds avdp.
Tank	Bombay (Mumbai)	68.1 grains (72 tanks=1 seer)
Teng	Burma	Burmese measures of capacity depend on the teng or basket; officially a basket is 2,218.2 cubic inches, but the teng varies locally: Akyab = 23 pounds of rice Bassein = 51 pounds of rice Moulmein = 48 pounds of rice Rangoon = 48 - 50 pounds of rice
Thanan or fannan	Thailand	1 litre
Thangsat	Thailand	21.3 litres
To	Japan	18.0391 litres=4.77 U.S. gallons or 2.05 U.S. pecks
Toise	France	1.949 metres
Tola	India	180 grains; legal weight of a silver rupee
Tomand	Arabian Peninsula	187.17 pounds avdp of rice
Tönde	Denmark	131.4 litres (liquid measure); 139.1 litres (dry measure)
Tonelada	Portugal	793.15 kilograms
Tonos	Greece	3,307 pounds avdp.
Tou	China	10 litres
Tovar	Bulgaria	128.2 kilograms

Unit	*Country/Place*	*Equivalent*
T'sun	China	1.41 inches
Tu	China	100.142 miles=250 li
Vara	Mexico	83.8 centimetres (32.99 inches)
	Chile, Peru and Spain	83.59 centimetres (32.91 inches)
	Argentina	86.6 centimetres (34.09 inches)
	Portugal	1.1 metres (43.31 inches)
	Texas (U.S.)	33.33 inches (84.67 centimetres)
Vat	The Netherlands	1 hectolitre; 768 mingelen (old value); 1 mingelen=1.20 to 1.237 litres
Vedro	U.S.S.R.	10 schtoffs, or 12.3 litres
	Bulgaria	10 litres
Verchok	U.S.S.R.	1.75 inches
Versta or verst	U.S.S.R.	0.662 88 mile
Vierkante roedo	The Netherlands	1 are
Viertel	Denmark	7.73 litres
	Switzerland	15 litres
Viss	Burma	3.60 pounds avdp (3.65 pounds avdp old value)
Wa or wah	Thailand	2 metres
Wichtje	The Netherlands	1 gram
Wisse	The Netherlands	1 stere
Zak	The Netherlands	1 hectolitre
Zar (metric)	Iran	1 metre
Zer	Iran	(*see Guz*)
Zoll	Switzerland	3 centimetres; old zoll nearly 1 inch
Zolotnik	U.S.S.R.	65.8306 grains, or 96 doli

5

Conversion Factors

A

To Convert	*into*	*multiply by*
Abampere *(electric current, charge/time)*	ampere(*A*)	10.0
Abampere	statampere	2.998×10^{10}
Abampere turn *(magnetic potential)*	ampere turn	10.0
Abampere turn	gilbert (*Gb*)	12.57
Abcoulomb (*dielectric flux*)	ampere-hour (*A.h*)	2.778×10^{-4}
Abcoulomb	coulomb (*C*)	10.0
Abcoulomb	faraday	1.036×10^{-4}
Abcoulomb	statcoulomb	2.998×10^{10}
Abfarad (*capacitance*)	farad (*F*)	1.0×10^{9}
Abfarad	statfarad	8.988×10^{20}
Abhenrie (*inductance*)	henrie (*H*)	1.0×10^{-9}
Abhenrie	stathenrie	1.112×10^{-21}
Abmho/cm (*electric conductivity-volumetric*)	mho (*mil-ft*)	1.662×10^{2}
Abmho/cm	mho(*m-g*)	$10^5/\delta$* (*mass density*)
Ahohm (*electric resistance*)	ohm (*Ω*)	1.0×10^{-9}
Abohm	statohm	1.112×10^{-21}

* *δ = density in g/cm³*

To Convert	*into*	*multiply by*
Abohm-cm (*electric resistivity-volume basis*)	microhm-cm	1.0×10^{-3}
Ahohm-cm	microhm-in.	3.937×10^{-4}
Abohm-cm	ohm (*mil-ft*)	6.015×10^{-3}
Abohm-cm	ohm (*m-g*) (*mass electric resistivity*)	$1.0 \times 10^{-5}\delta$*
Abvolt (*electromotive force*)	statvolt	3.335×10^{-11}
Abvolt	volt (*V*)	1.0×10^{-8}
Acre (*area*)	are (*area*)	40.469 45
Acre	bigha	1.6
Acre	chain^2 (*Gunter's*)	10.0
Acre	ft^2	4.356×10^4
Acre	hectare or hectometre^2	0.404 686
Acre	in.^{2}	$6.272\ 64 \times 10^6$
Acre	km^2	4.047×10^{-3}
Acre	link^2 (*Gunter's*)	1.0×10^5
Acre	m^2	$4.046\ 856 \times 10^3$
Acre	mile^2	$1.562\ 5 \times 10^{-3}$
Acre†	rod^2	1.6×10^2
Acre	rood	4.0
Acre	yard^2	4.84×10^3
Acre-ft (*volume, irrigation, reservoir*)	ft^3	4.356×10^4
Acre-ft	kilolitre	$1.233\ 441\ 1 \times 10^3$
Acre-ft	gallon (*U.K.*)	$2.713\ 35 \times 10^5$
Acre-ft	gallon (*U.S.*)	$3.258\ 5 \times 10^5$
Acre-ft	m^3	$1.233\ 482 \times 10^3$

* *δ = density in g/cm³*

† *1 acre (English system) = 4 rods × 40 rods; 1 rod = 5½ yards or 16½ ft, perch is its earlier name*

To Convert	*into*	*multiply by*
Acre-ft	$yard^3$	$1.613\ 345\ 6 \times 10^3$
Ampere(*A*) (*electric current, charge/time*)	abampere	0.1
Ampere (*absolute*)	ampere (*International*)	1.0002
Ampere (*International*)	ampere (*absolute*)	0.999 8
Ampere	cgse*	3.0×10^9
Ampere	cgsm*	0.1
Ampere	faraday/sec	$1.036\ 269\ 4 \times 10^{-5}$
Ampere	statampere	2.998×10^9
$Ampere/cm^2$ (*density, current*)	ampere/$in.^2$	6.452
$Ampere/cm^2$	$ampere/m^2$	1.0×10^4
$Ampere/ft^2$	$ampere/m^2$	10.763 91
Ampere-hour (*A.h*)	abcoulomb	3.6×10^2
Ampere-hour	coulomb	3.6×10^3
Ampere-hour	faraday	0.037 31
Ampere-hour	statcoulomb	1.08×10^{13}
Ampere/in. (*magnetization, coil field strength*)	ampere/m	39.370 08
Ampere/$in.^2$ (*density, current*)	$ampere/cm^2$	0.154 990 7
Ampere/$in.^2$	kA/m^2	1.550 003
Ampere/m	ampere/in.	$2.539\ 48 \times 10^{-2}$
Ampere/m	cgse	3.77×10^8
Ampere/m (*magnetic field intensity*)	oersted(*Oe*)	1.26×10^{-2}
$Ampere/m^2$	$ampere/cm^2$	1.0×10^{-4}
$Ampere/m^2$	$ampere/ft^2$	9.2903×10^{-2}
$Ampere/m^2$	ampere/$in.^2$	6.452×10^{-4}

* *Symmetrical or Gaussian electrostatic system of units*

To Convert	*into*	*multiply by*
Ampere-turn (*magnetic potential*)	abampere-turn	0.1
Ampere-turn	cgse	3.77×10^{10}
Ampere-turn (*magneto motive force*)	gilbert (*Gb*)	1.257
Ampere-turn/cm (*magnetic field intensity*)	ampere-turn/in.	2.54
Ampere-turn/cm	gilbert/cm	1.257
Ampere-turn/in.	ampere-turn/cm	0.393 7
Ampere-turn/in.	gilbert/cm	0.495
Ampere-turn/m	gilbert/cm	1.257×10^{-2}
Angstrom unit (Å) (*wave length*)	in.	3.937×10^{-9}
Angstrom unit	m	1.0×10^{-10}
Angstrom unit	micron (μ)	1.0×10^{-4}
Angstrom unit	nanometre	10.0
Apostilb (*luminance*, $\pi^{-1} cd/m^2$)	lambert(*L*)	1.0×10^{-4}
Apostilb	nit	$1/\pi$
Apostilb	stilb	3.18×10^{-5}
Arc (*angular, earth*)	hour(*h*)	$0.06\overline{6}$*
Arc	minute(*time*)	4.0
Arc	sec (*time*)	2.4×10^{2}
Are (*area*)	acre(*US*)	2.471×10^{-2}
Are	m^2	1.0×10^{2}
Are	$yard^2$	1.196×10^{2}
Astronomical unit	km	1.495×10^{8}
Atmosphere (*pressure*)	bar	1.013 273 8
Atmosphere	millibar	$1.013\ 253\ 354 \times 10^{3}$

** symbol — over a digit indicate recurring decimal digit*

To Convert	*into*	*multiply by*
Atmosphere	cm of Hg	76.0
Atmosphere	dyne/cm^2	1.013 273 8 × 10^6
Atmosphere	ft of water(*at 4°C*)	33.9
Atmosphere	in. of mercury (*at 0°C*)	29.921 3
Atmosphere	in. of water (*at 4°C*)	4.067 818 673 × 10^2
Atmosphere	kg/cm^2	1.0333
Atmosphere	kg/m^2	1.033 228 63 × 10^4
Atmosphere	lb/in.^{2}	14.7
Atmosphere	metre of water	10.34
Atmosphere	metre-kgf/sec	10.341 261 634
Atmosphere	mm of mercury	7.6 × 10^2
Atmosphere (*normal: 760 torr pressure*)	pascal(*Pa*)	1.013 25 × 10^5
Atmosphere	pieze	1.01 × 10^2
Atmosphere	pound/ft^2	2.116 218 49 × 10^3
Atmosphere	tonne/ft^2	1.058
Atmosphere	tonne/in.^{2}	7.348 × 10^{-3}
Atomic mass unit (*energy*)	British thermal unit (*Btu*)	1.414 × 10^{-13}
Atomic mass unit	calorie	3.565 × 10^{-11}
Atomic mass unit	erg	1.492 × 10^{-3}
Atomic mass unit	ft-lb	1.101 × 10^{-10}
Atomic mass unit	ft-poundal	3.541 × 10^{-9}
Atomic mass unit	joule	1.492 × 10^{-10}
Atomic mass unit	kWh	4.145 × 10^{-17}

B

To Convert	*into*	*multiply by*
Bar* (*pressure*)	atmosphere	0.9869
Bar	dyne/cm^2	1.0×10^6
Bar	kgf/m^2	$1.019\ 72 \times 10^4$
Bar	lbf/ft^2	$2.088\ 54 \times 10^3$
Bar	lbf/in.^{2}	14.5038
Bar	Pa	1.0×10^5
Bar	pieze (*pz*)	1.0×10^2
Bar	poundal/ft^2	6.7197×10^4
Bar	tonne f/in.^{2}	6.4749×10^{-3}
Barn (*cross sectional area*)	m^2	1.0×10^{-28}
Barrel	hogshead	0.5
Barrel (*oil*)	gallon (*oil*)	42.0
Barrel (*US*)	bushel (*UK*)	4.371 57
Barrel (*US*)	bushel (*US*)	4.511 68
Barrel (*US; dry*)	quart (*dry*)	1.05×10^2
Barrel (*US; dry*)	in.^{3}	7.056×10^3
Barrel (*US; liquid*)	ft^3	5.6146
Barrel (*US; liquid*)	gallon	31.5
Barrel (*US; liquid*)	kilolitre	0.158 984
Barrel (*US; liquid*)	m^3	0.158 988
Barrel (*US; liquid*)	yard^3	0.207 949
Baryl	dyne/cm^2	1.0
Beequerel	curie (*radio-activity*)	$2.702\ 702\ 7 \times 10^{-11}$
Bigha	acre	0.625
Bigha	metre^2	$2.529\ 285\ 264 \times 10^3$
Bigha	yard^2	3.025×10^3
Bolt (*US; cloth*)	m (*metre*)	36.576

* *1 Bar = 10^5 newton/m^2*

To Convert	*into*	*multiply by*
Btu	atomic mass unit	7.07×10^{12}
Btu	erg	1.055×10^{10}
Btu	ft-lb	7.7826×10^{2}
Btu	gram-calorie	$2.519\ 96 \times 10^{2}$
Btu	gram-cm	$1.075\ 615\ 79 \times 10^{7}$
Btu	hoursepower-hour	$3.930\ 15 \times 10^{-4}$
Btu	in.-lb	$9.339\ 12 \times 10^{3}$
Btu	joule (*J*)*	$1.055\ 056 \times 10^{3}$
Btu (*mean*)	joule	$1.055\ 87 \times 10^{3}$
Btu	kg-calorie	0.252 016 129
Btu	kg-m	1.075×10^{2}
Btu	kW-hour	$2.930\ 71 \times 10^{-4}$
Btu	litre-atmosphere	10.409
Btu	watt-hour	0.292 997 363
Btu/^{0}F	kJ/K	1.899 101
Btu/ft^2(*solar radiant exposure*)	J/m^2	$1.135\ 652\ 8 \times 10^{4}$
Btu/ft^3 (*calorific value*)	kJ/m^3	37.26
Btu/ft^3-^{0}F (*specific heat, volume basis*)	J/cm^3-^{0}C	6.7066×10^{-2}
Btu/ft^3-^{0}F	kcal/m^3-^{0}C	16.0185
Btu/ft^2-h	cal/cm^2-s	7.5346×10^{-5}
Btu/ft^2-h	kcal/m^2-h	2.712 46
Btu/ft^2-h (*solar radiation*)	W/m^2	3.154 591
Btu/ft-h-^{0}F	Btu-in./ft^2-h-^{0}F	12.0
Btu/ft-h-^{0}F	cal/cm-s-^{0}C	$4.133\ 79 \times 10^{-3}$
Btu/ft-h-^{0}F	kcal/m-h-^{0}C	1.488 16
Btu/ft-h-^{0}F	W/cm-^{0}C	$1.730\ 73 \times 10^{-2}$

* *Overall energy production and consumption is also reported in Q's (10^{18} Btu) and Quad or quadrillion (10^{15} Btu)*

To Convert	*into*	*multiply by*
Btu/ft^2-h-°F (*thermal conductance*)	cal/cm^2-s-°C	$1.356\,23 \times 10^{-4}$
Btu/ft^2-h-°F	kcal/m^2-h-°C	4.882 43
Btu/ft^2-h-°F	W/cm^2-°C	$5.67\,83 \times 10^{-4}$
Btu/ft^2-h-F°/ft (*thermal conductance*)	in.-lb/in.^{2}-s-°F/in.	0.216
Btu-ft/(h-ft^2-°F)	W/(m-K)	1.730 735
Btu/ft/min (*solar irradiance, insolation*)	W/in.^{2}	0.1221
Btu/ft^2/min	W/m^2	$1.891\,489 \times 10^2$
Btu/gallon (*calorific value*)	kJ/litre	0.2321
Btu/h	cal/sec	$6.\overline{9}^* \times 10^{-2}$
Btu/h	ft-lb/sec	0.2162
Btu/h	gram-cal/sec	7.0×10^{-2}
Btu/h	horsepower	3.929×10^{-4}
Btu/h	kcal/h	0.251 996
Btu/h	W	0.293 071
Btu/(h-ft^2) (*density of heat flow rate, irradiance*)	W/m^2	3.154 591
Btu/(h-ft^2-°F)	W/(m^2-K)	5.678 263
Btu/in.^{3}-°F (*volumetric specific heat*)	in.-lb/in.^{3} °F	9.33×10^3
Btu-in./ft^2-h-°F	Btu/ft-h-°F	$8.3\overline{3}^* \times 10^{-2}$
Btu-in./ft^2-h-°F	cal/cm-s-°C	$3.444\,82 \times 10^{-4}$
Btu-in./ft^2-h-°F	kcal/m-h-°C	0.124 014
Btu-in./ft^2-h-°F	W/cm-°C	$1.442\,28 \times 10^{-3}$
Btu-in./ft^2-h-°F	W/m-K	0.1441
Btu/lb (*calorific value*)	joule/gram	2.325 997 911
Btu/lb	kcal/kg	$0.55\overline{5}^*$
Btu/lb	kJ/kg	2.326
Btu/lb-°F (*specific heat, mass basis*)	cal/gram-°C	1.0

* *symbol — over a digit indicate recurring decimal digit*

To Convert	*into*	*multiply by*
Btu/lb-^{0}F	J/gram-^{0}C	4.1868
Btu/min	erg/sec	$1.758\ 087\ 2 \times 10^{8}$
Btu/min	ft-lb/min	7.7826×10^{2}
Btu/min	ft-lb/sec	12.971
Btu/min	horsepower (*hp*)	2.356×10^{-2}
Btu/min	horsepower (*metric*)	2.39×10^{-2}
Btu/min	kW	$1.757\ 96 \times 10^{-2}$
Btu (*mean*)/min	watt (*absolute*)	17.579 635 75
Btu/ft^2-hour	watt/cm^2	$3.154\ 594 \times 10^{-4}$
Bucket (*UK;dry*)	cm^3	1.818×10^{4}
Bushel	chaldron	$2.7\overline{7}^{*} \times 10^{-2}$
Bushel	cm^3	3.524×10^{4}
Bushel	in.^{3}	$2.150\ 42 \times 10^{3}$
Bushel	peck	4.0
Bushel	pint (*dry*)	64.0
Bushel	quart (*dry*)	32.0
Bushel	stere (*volume*)	$3.524\ 229\ 07 \times 10^{-2}$
Bushel	tonne (*shipping, UK, cargo*)	$3.057\ 823\ 44 \times 10^{-2}$
Bushel	tonne (*shipping, US, cargo*)	$3.111\ 097\ 28 \times 10^{-2}$
Bushel (*combine harvester grain tank capacity*)	litre	35.239 07
Bushel (*US*)	barrel (*US*)	0.221 647
Bushel (*US*)	bushel (*UK*)	0.968 95
Bushel (*US*)	ft^3	1.244 46
Bushel (*US*)	kilolitre	$3.523\ 83 \times 10^{-2}$
Bushel (*US*)	yard^3	$4.609\ 12 \times 10^{-2}$
Bushel (*US; volume, grain bin*)	m^3	$3.523\ 907 \times 10^{-2}$

* *symbol — over a digit indicate recurring decimal digit*

To Convert	*into*	*multiply by*
Bushel (*UK*)	barrel (*US*)	0.228 751
Bushel (*UK*)	bushel (*US*)	1.032 05
Bushel (*UK*)	ft^3	1.284 35
Bushel (*UK*)	kilolitre	$3.636\ 768 \times 10^{-2}$
Bushel (*UK*)	m^3	$3.636\ 87 \times 10^{-2}$
Bushel (*UK*)	$yard^3$	$4.756\ 85 \times 10^{-2}$
Bushel(60 lb)/acre (*crop yield*)	kg/ha	67.26

C

To Convert	*into*	*multiply by*
Cable length	fathom	1.2×10^2
Cable length	furlong	1.090 91
Cable length	kilometre	0.219 456
Cable length	mile (*nautical; International*)	0.118 497
Cable length	mile (*nautical; UK*)	0.118 421
Cable length	mile (*statute*)	0.136 364
Cable length	yard	2.4×10^2
Calorie	atomic mass unit	2.805×10^{10}
Calorie	erg	4.1868×10^7
Calorie	in.-pound	37.1
Calorie (*mean*)	joule	4.190 02
Calorie	kgf-metre	0.427
Calorie*	kWh	1.163×10^{-6}
Calorie/cm-s-°C	Btu/ft-h-°F	$2.419\ 09 \times 10^2$
Calorie/cm-s-°C	Btu-in./ft²-h-°F	$2.902\ 91 \times 10^3$
Calorie/cm-s-°C	kcal/m-h-°C	3.6×10^2
Calorie/cm-s-°C	W/cm-°C	4.1868
Calorie/cm² -s	Btu/ft² -h	$1.327\ 21 \times 10^4$
Calorie/cm² -s	kcal/m² -h	3.6×10^4
Calorie/cm² -s	W/cm²	4.183 999 8
Calorie/cm² -s-°C (*thermal conductance*)	Btu/ft² -h-°F	7.3734×10^3
Calorie/cm² -s-°C	kcal/m² -h-°C	3.6×10^4
Calorie/cm² -s-°C	W/cm² - °C	4.1868
Calorie/cm² -s-°C/cm (*thermal conductivity*)	pound**-s-°F	52.2

* *Calorie (diet) = 1.163 × 10⁻³ kWh is also used*

** *pound denotes pound mass*

To Convert	*into*	*multiply by*
Calorie/cm³ -°C (*volumetric specific heat*)	pound/in.² - °F	3.38×10^2
Calorie/gram (*heat, specific*)	joule/kg	4.1868×10^3
Calorie/gram-°C (*specific heat*)	Btu/lb- F (*specific heat, mass basis*)	1.0
Calorie/gram-°C (*specific heat*)	joule/g-°C	4.1868
Calorie/sec	Btu/hour	14.286
Calorie/sec	kcal/hour	3.6
Calorie/sec	W	4.1868
Candle/cm²	lambert(*L*)	3.142
Candle/in.²	lambert	0.487
Carat (*metric*)	drachm (*apothacaries*)	5.1441×10^{-2}
Carat (*metric*)	dram (*avoirudupois*)	0.112 877
Carat (*metric*)	grain	3.086 47
Carat (*metric*)	gram (*g*)	0.2
Carat (*metric*)	ounce (*apothacaries*)	6.4301×10^{-3}
Carat (*metric*)	ounce (*avoirdupois*)	7.0548×10^{-3}
Carat (*metric*)	tola*	$1.714\ 71 \times 10^{-2}$
Cent	savart	0.251
Cent (*natural*)	fifth (*G-tone, musical interval*)	1.4245×10^{-3}
Cent (*tempered*)	fifth (*G-tone, musical interval*)	$1.428\ 57 \times 10^{-3}$
Cent (*natural*)	fourth (*F-tone, musical interval*)	2.008×10^{-3}
Cent (*tempered*)	fourth (*F-tone,*	2.0×10^{-3}

***** *Indian weight measure for small quantities (gold, silver etc.)***

To Convert	*into*	*multiply by*
	musical interval)	
Cent (*natural*)	musical interval (*D-tone, major*)	$4.901\ 96 \times 10^{-3}$
Cent (*tempered*)	musical interval (*D-tone, major*)	5.0×10^{-3}
Cent (*natural*)	octave (*C-tone, musical interval*)	8.33×10^{-4}
Cent (*tempered*)	octave (*C-tone, musical interval*)	8.33×10^{-4}
Cent (*natural*)	Seventh (*HCB-tone*)	9.1912×10^{-4}
Cent (*tempered*)	Seventh (*HCB-tone*)	9.0909×10^{-4}
Cent (*natural*)	Sixth (*A-tone, major*)	$1.131\ 22 \times 10^{-3}$
Cent (*tempered*)	Sixth (*A-tone, major*)	$1.111\ 11 \times 10^{-3}$
Cent (*natural*)	Third (*E-tone, major*)	$2.590\ 67 \times 10^{-3}$
Cent (*tempered*)	Third (*E-tone, major*)	2.5×10^{-3}
Cent (*natural*)	Unison (*C-tone, musical interval*)	$1.666\ 67 \times 10^{-3}$
Cent (*tempered*)	Unison (*C-tone, musical interval*)	$1.666\ 67 \times 10^{-3}$
Centare (*Centiare*)	m^2	1.0
Celsius (°C)	fahrenheit (°F)	°F=(°C × 9/5 + 32)
Celsius (°C)	kelvin (K)	K= °C+273.15
Centigram	gram (*g*)	1.0×10^{-2}
Centilitre	dram	2.705
Centilitre	gill (*liquid*)	8.45×10^{-2}
Centilitre	$in.^3$	0.6103
Centilitre	litre	1.0×10^{-2}

To Convert	*into*	*multiply by*
Centilitre	ounce (*fluid; US*)	0.3382
Centimetre (*cm*)	ell	$8.748\ 91 \times 10^{-3}$
Centimetre	em, pica	2.362 390 74
Centimetre	feet (*ft*)	3.281×10^{-2}
Centimetre	hand	$9.842\ 52 \times 10^{-2}$
Centimetre	inch (*in.*)	0.393 700 7874
Centimetre	kilometre (*km*)	1.0×10^{-5}
Centimetre	metre (*m*)	1.0×10^{-2}
Centimetre	mil	3.937×10^{2}
Centimetre	mile	6.214×10^{-6}
Centimetre	mile (*nautical*)	$5.396\ 654\ 1 \times 10^{-6}$
Centimetre	millimetre (*mm*)	10.0
Centimetre	yard	1.094×10^{-2}
Centimetre^2	circular mil	1.973×10^{5}
Centimetre^2	ft ^{2}	1.076×10^{-3}
Centimetre^2	in.^{2}	0.155 000 31
Centimetre^2	kilometre^2	1.0×10^{-10}
Centimetre^2	metre^2	1.0×10^{-4}
Centimetre^2	mil^2	$1.549\ 907 \times 10^{5}$
Centimetre^2	mile^2	3.861×10^{-11}
Centimetre^2	millimetre^2	1.0×10^{2}
Centimetre^2	yard^2	1.196×10^{-4}
Centimetre^3	bucket (*UK, dry*)	$5.500\ 550\ 1 \times 10^{-5}$
Centimetre^3	bushel	$2.837\ 684\ 4 \times 10^{-5}$
Centimetre^3	dram (*US fluid or apothecarie's*)	0.270 511 54
Centimetre^3	ft ^{3}	3.531×10^{-5}
Centimetre^3	gallon (*liquid; US*)	2.642×10^{-4}
Centimetre^3	gill (*UK*)	7.039×10^{-3}
Centimetre^3	in.^{3}	$6.102\ 374\ 409 \times 10^{-2}$

To Convert	*into*	*multiply by*
Centimetre^3	litre	$9.999\,72 \times 10^{-4}$
Centimetre^3	metre^3	1.0×10^{-6}
Centimetre^3	millilitre	0.999 97
Centimetre^3	minim (*UK*)	16.8936
Centimetre^3	minim (*US*)	16.2306
Centimetre^3	ounce (*UK*)	$3.519\,51 \times 10^{-2}$
Centimetre^3	ounce (*US*)	$3.381\,38 \times 10^{-2}$
Centimetre^3	pint (*US, liquid*)	2.113×10^{-3}
Centimetre^3	quart (*US, liquid*)	1.057×10^{-3}
Centimetre^3	yard^3	1.308×10^{-6}
Centimetre^3/sec	ft^3/minute	$2.118\,88 \times 10^{-3}$
cm-dyne	cm-gram	1.02×10^{-3}
cm-dyne	ft-pound	7.376×10^{-8}
cm-dyne	metre-kg	1.02×10^{-8}
cm-gram	cm-dyne	9.807×10^{2}
cm-gram	erg	$9.803\,921\,6 \times 10^{2}$
cm-gram	ft-pound	7.233×10^{-5}
cm-gram	metre-kg	1.0×10^{-5}
cm of mercury	atmosphere	1.316×10^{-2}
cm of mercury	feet of water	0.4461
cm of mercury	kg/metre^2	1.36×10^{2}
cm of mercury (*at ^0C*)	pascal	1.333×10^{3}
cm of mercury	pound/ft^2	27.85
cm of mercury	pound/in.^{2}	0.1934
cm of water (*at 4 ^0C*)	pascal	98.06
cm/sec	feet/minute	1.1969
cm/sec	feet/sec	3.281×10^{-2}
cm/sec	km/hour	3.6×10^{-2}
cm/sec	knot	1.943×10^{-2}
cm/sec	metre/minute	0.6
cm/sec	mile/hour	2.237×10^{-2}

To Convert	*into*	*multiply by*
cm/sec	mile/minute	3.728×10^{-4}
cm/sec/sec	ft/sec/sec	3.281×10^{-2}
cm/sec/sec	km/hour/sec	3.6×10^{-2}
cm/sec/sec	metre/sec/sec	1.0×10^{-2}
cm/sec/sec	mile/hour/sec	2.237×10^{-2}
cm^2/sec (*thermal diffusivity*)	ft^2/h	3.873 708 4
cm^3/dyne-sec (*acoustic admittance, acoustic conductance and acoustic susceptance*)	$metre^3$/(Pa-sec)	1.0×10^{-5}
cm^4	$in.^4$	$2.402\ 51 \times 10^{-2}$
Centipoise (*viscosity, dynamic*)	pascal-sec	1.0×10^{-3}
Centipoise	pound force-sec/ft^2	2.089×10^{-5}
Centipoise	pound mass/ft-hour	2.42
Centipoise	pound mass/ft-sec	6.72×10^{-4}
Centistoke (*viscosity, kinematic*)	$metre^2$/sec	1.0×10^{-6}
Chain (*surveyor's or Gunter's*)	fathom	11.0
Chain	feet	66.0
Chain	furlong	0.1
Chain	in.	7.92×10^2
Chain	link	1.0×10^2
Chain	metre	20.1168
Chain	rod	4.0
Chain	yard	22.0
$Chain^2$ (*Gunter's*)	acre (*area*)	0.1
$Chain^2$	ft^2	4.356×10^3
$Chain^2$	rod^2	16.0
$Chain^2$	rood	0.4

To Convert	*into*	*multiply by*
$Chain^2$	$yard^2$	4.84×10^{2}
cgse* (*charge*)	cgsm**	$3.3\overline{3}$*** $\times 10^{-11}$
cgse	coulomb (*C*)	$3.3\overline{3}$*** $\times 10^{-10}$
cgsm (*charge*)	cgse	3.0×10^{10}
cgsm	coulomb (*C*)	10.0
cgse (*field intensity*)	cgsm	3.0×10^{10}
cgse	volt/metre	3.0×10^{4}
cgsm	cgse	$3.3\overline{3}$*** $\times 10^{-11}$
cgsm	volt/metre	1.0×10^{-6}
cgse (*capacitance*)	cgsm	1.11×10^{-21}
cgse	farad (*F*)	1.11×10^{-12}
cgsm	cgse	8.99×10^{20}
cgsm	farad (*F*)	1.0×10^{9}
cgse (*current intensity*)	ampere (*A*)	$3.3\overline{3}$*** $\times 10^{-10}$
cgse	ampere-turn	2.65×10^{-11}
cgse	cgsm	$3.3\overline{3}$*** $\times 10^{-11}$
cgsm	ampere (*A*)	10.0
cgsm	cgse	3.0×10^{10}
Charge (*displacement flux*)	cgse	3.77×10^{10}
Charge (*displacement flux*)	cgsm	1.26
cgse	cgsm	$3.3\overline{3}$*** $\times 10^{-11}$
cgse	Charge (*displacement flux*)	2.65×10^{-11}
cgsm	cgse	3.0×10^{10}
cgsm	Charge (*displacement flux*)	0.796
cgse (*induction*)	cgsm	8.99×10^{20}
cgse	henry (*H*)	8.99×10^{11}

* *cgse : electrostatic system also known as electrostatic unit (esu).*

** *cgsm: electro magnetic system, also known as electro-magnetic unit (emu)*

*** *symbol — over a digit indicate recurring decimal digit*

To Convert	*into*	*multiply by*
cgsm	cgse	1.11×10^{-21}
cgsm	henry (*H*)	1.0×10^{-9}
cgse (*magnetic induction*)	gauss (*Gs*)	3.0×10^{10}
cgse	tesla (*T*)	3.0×10^{6}
cgse (*magnetic flux*)	maxwell (*Mx*)	3.0×10^{10}
cgse	weber (*Wb*)	3.0×10^{2}
cgse (*potential difference*)	cgsm	3.0×10^{10}
cgse	volt	3.0×10^{2}
cgsm (*potential difference*)	cgse	$3.3\overline{3}^{*} \times 10^{-11}$
cgsm	volt	1.0×10^{-8}
cgse (*resistance, electric*)	cgsm	8.99×10^{20}
cgse	ohm (*Ω*)	8.99×10^{11}
cgsm	cgse	1.11×10^{-21}
cgsm	ohm (*Ω*)	1.0×10^{-9}
cgse (*resistivity, electric*)	cgsm	8.99×10^{20}
cgse	ohm-metre (*Ω-m*)	8.99×10^{9}
cgse	ohm-mm²/metre	8.99×10^{15}
cgsm	cgse	1.11×10^{-21}
cgsm	ohm-metre (*Ω-m*)	1.0×10^{-11}
cgsm	ohm-mm²/metre	1.0×10^{-5}
cgse (*magnetic field intensity*)	ampere/metre	2.65×10^{-9}
cgse	oersted (*Oe*)	$3.3\overline{3}^{*} \times 10^{-11}$
cgse (*magnetomotive force*)	gilbert (*Gb*)	$3.3\overline{3}^{*} \times 10^{-11}$
cgse (*displacement*)	cgsm	$3.3\overline{3}^{*} \times 10^{-11}$
cgse	charge/metre²	2.65×10^{-7}
Charge/m²	cgse	3.77×10^{6}
Charge/m²	cgsm	1.26×10^{-4}
cgsm	cgse	3.0×10^{10}
cgsm	charge/metre²	7.96×10^{3}

***** *symbol — over a digit indicate recurring decimal digit***

To Convert	*into*	*multiply by*
Charge/m² (*surface charge density*)	cgse	3.0×10^{5}
Charge/m²	cgsm	1.0×10^{-5}
cgse (*cgs*)	cgsm	$3.3\overline{3}^* \times 10^{-11}$
cgse	charge/m²	$3.3\overline{3}^* \times 10^{-6}$
cgsm	cgse	3.0×10^{10}
cgsm	charge/m²	1.0×10^{5}
Charge/m³ (*volume density charge*)	cgse	3.0×10^{3}
Charge/m³	cgsm	1.0×10^{-7}
cgse (*volume density charge*)	cgsm	$3.3\overline{3}^* \times 10^{-11}$
cgse	charge/m³	$3.3\overline{3}^* \times 10^{-4}$
cgsm	cgse	3.0×10^{10}
cgsm	charge/m³	1.0×10^{7}
Chaldron	bushel	36.0
Charge (C)	cgse	3.0×10^{9}
cgse	charge	$3.3\overline{3}^* \times 10^{-10}$
Charge	cgsm	0.1
cgsm	charge	10.0
Chemical mass unit	kg	$1.660\ 24 \times 10^{-27}$
Cheval vapeur	ft-poundf/sec	5.4248×10^{2}
Cheval vapeur	hp	0.986 32
Cheval vapeur	kilowatt	0.7355
Cheval vapeur	metre-kgf/sec	75.0
Circular mil**	cm²	5.067×10^{-6}
Circular mil	ft²	$5.455\ 537\ 4 \times 10^{-9}$
Circular mil	in.²	7.854×10^{-7}
Circular mil	metre²	$5.068\ 423 \times 10^{-10}$
Circular mil	mil²	0.7854
Circular mil	mm²	5.067×10^{-4}

** symbol — over a digit indicate recurring decimal digit*

*** circular mil is the area of a circle, 1 mil in diameter and circular inch is the area of a circle 1 inch in diameter*

To Convert	*into*	*multiply by*
Circumference	radian	6.283
Cord	cord ft	8.0
Cord*	ft^3	1.28×10^2
Cord feet	cord	0.125
Cord feet	ft^3	16.0
Coulomb	abcoulomb	0.1
Coulomb	ampere-hour	2.778×10^{-4}
Coulomb	cgse	3.0×10^9
Coulomb	cgsm	0.1
Coulomb	faraday	1.036×10^{-5}
Coulomb	statcoulomb	2.998×10^9
Coulomb/cm^2	coulomb/$in.^2$	6.452
Coulomb/cm^2	coulomb/$metre^2$	1.0×10^4
Coulomb/$in.^2$	coulomb/cm^2	0.155
Coulomb/$in.^2$	coulomb/$metre^2$	1.55×10^3
Coulomb/$metre^2$	coulomb/$in.^2$	6.452×10^{-4}
Coulomb/$metre^2$	Coulomb/cm^2	1.0×10^{-4}
Cubit	in.	18.0
Cubit (*Bible*)	in.	21.8
Cup	gallon (*US; liquid*)	6.25×10^{-2}
Cup	quart (*fluid*)	0.25
Curie (*radio-activity*)	beequerel	3.7×10^{10}
Cwt (*mass*)	kg	50.802
Cwt (*mass*)	pound	1.12×10^2
Cwt	pound (*troy*)	$1.361\,118\,3 \times 10^2$
Cwt (*avoirdupois*)	quarter (*avoirdupois*)	4.0
Cwt (*mass*)	tonne (*long*)	5.0×10^{-2}
Cwt (*mass*)	tonne (*short*)	5.6×10^{-2}
Cycles/sec	hertz (*Hz*)	1.0

* *1 cord of wood equals to a pile cut 4 ft long piled 4 ft high and 8 ft on the ground equals to 128 ft^3*

D

To Convert	*into*	*multiply by*
Dalton	gram	1.65×10^{-24}
Day	hour	24.0
Day (*mean solar*)	secondi	8.64×10^{4}
Decibel*	neper	0.115 127 79
Decigram	gram	0.1
Decilitre	litre	0.1
Decimetre	metre	0.1
Degree (*angle*)	gon	1.11
Degree	metric minute (c)	1.11×10^{2}
Degree	metric second (cc)	1.11×10^{4}
Degree	mil	$17.7\overline{7}$**
Degree	minute	60.0
Degree	quadrant	$1.1\overline{1}$** $\times 10^{-2}$
Degree	radian	$1.745\ 33 \times 10^{-2}$
Degree	revolution	$2.7\overline{7}$** $\times 10^{-3}$
Degree	second	3.6×10^{3}
Degree	sign	$3.\overline{3}$** $\times 10^{-2}$
Degree/sec	radian/sec	$1.745\ 33 \times 10^{-2}$
Degree/sec	revolution/minute	$0.16\overline{6}$**
Degree/sec	revolution/sec	$2.7\overline{7}$** $\times 10^{-3}$
Dekagram	gram	10.0
Dekalitre	litre	10.0
Dekametre	metre	10.0
Dekametre	rod	1.9884

** Whenever the intensity of sound increases by a factor of 10, the increase in intensity is said to be 1 bel or 10 decibel*

*** symbol — over a digit indicate recurring decimal digit*

To Convert	*into*	*multiply by*
Drachm (*apothecaries' or troy*)	carat (*metric*)	19.4397
Drachm (*apothecaries' or troy*)	dram (*avoirdupois*)	2.194 29
Drachm	grain	60.0
Drachm	gram	3.887 93
Drachm	ounce (*avoirdupois*)	0.137 142 9
Drachm	ounce (*troy*)	0.125
Drachm (*apothecaries' or troy*)	pound (*avoirdupois*)	$8.571\,43 \times 10^{-3}$
Drachm (*apothecaries' or troy*)	tola	$0.3\overline{3}$*
Dram (*avoirdupois*)	carat (*metric*)	8.8592
Dram	centilitre	0.369 685 77
Dram (*US fluid or apothecaries'*)	cm^3	3.6967
Dram	drachm (*apothecaries' or troy*)	0.455 729
Dram	grain	27.343 75
Dram	gram	1.771 85
Dram	ounce (*apothecaries' or troy*)	5.6966×10^{-2}
Dram	ounce (*avoir dupois*)	6.25×10^{-2}
Dram	pound (*avoirdupois*)	$3.906\,25 \times 10^{-3}$
Dram	scruple	3.0
Dram	tola	0.151 908 73
Dyne	gramf	1.02×10^{-3}

***** *symbol — over a digit indicate recurring decimal digit*

To Convert	*into*	*multiply by*
Dyne	joule/cm	1.0×10^{-7}
Dyne	joule/metre (*newton*)	1.0×10^{-5}
Dyne	kgf	1.02×10^{-6}
Dyne	newton	1.0×10^{-5}
Dyne	poundf (*lbf*)	2.248×10^{-6}
Dyne	pieze	1.0×10^{-4}
Dyne	poundal	7.233×10^{-5}
Dyne-cm	erg	1.0
Dyne/cm	erg/mm^2	1.0×10^{-2}
Dyne/cm (*tension, surface*)	mN/m	1.0
Dyne/cm^2 (*microbar*)	atmosphere	9.869×10^{-7}
Dyne/cm^2	bar	1.0×10^{-6}
Dyne/cm^2	baryl	1.0
Dyne/cm^2	in. of mercury at 0^0C	2.953×10^{-5}
Dyne/cm^2	in. of water at 4^0C	4.015×10^{-4}
Dyne/cm^2	kg/cm^2	$1.019\ 716\ 2 \times 10^{-6}$
Dyne/cm^2	lb/in.^{2}	14.5
Dyne/cm^2	mm of mercury	7.5×10^{-4}
Dyne/cm^2	ounce/in.^{2}	$2.320\ 724\ 1 \times 10^{-4}$
Dyne/cm^2	Pa	0.1
Dyne-cm/sec	erg/sec	1.0
Dyne-sec/cm^5 (*acoustic impedance/resistance/reactance*)	Pa-s/m	1.0×10^{5}

E

To Convert	*into*	*multiply by*
Einstein (*photon energy*)	quantum	$6.022\,045 \times 10^{23}$
Electronvolt (*energy*)	joule	$1.602\,19 \times 10^{-19}$
Ell	cm	1.143×10^{2}
Ell	in.	45.0
Em, Pica	in.	0.167
Em, Pica	cm	0.4233
Erg	atomic mass units	6.702×10^{2}
Erg (*energy*)	Btu	9.48×10^{-11}
Erg	calorie	$2.388\,459 \times 10^{-8}$
Erg	cm-gram	1.02×10^{-3}
Erg	dyne-cm	1.0
Erg	ft-lb	7.3756×10^{-8}
Erg	ft-poundal	2.373×10^{-8}
Erg	gram-calorie	2.389×10^{-8}
Erg	horsepower-hour	3.725×10^{-14}
Erg	joule	1.0×10^{-7}
Erg	kg-calorie	2.389×10^{-11}
Erg	kilowatt-hour	$2.7\overline{7}$* $\times 10^{-14}$
Erg	metre-kgf	1.02×10^{-8}
Erg	watt/metre^2	1.0×10^{-3}
Erg	watt-hour	$2.7\overline{7}$* $\times 10^{-11}$
Erg/cm^3 (*sound energy density*)	joule/metre^3	0.1
Erg/mm^2	dyne/cm	1.0×10^{2}
Erg-sec	planck's quantum	$1.509\,661\,8 \times 10^{26}$
Erg/sec	Btu/minute	5.688×10^{-9}
Erg/sec	dyne-cm/sec	1.0
Erg/sec	ft-lb/minute	4.4254×10^{-6}

** symbol — over a digit indicate recurring decimal digit*

To Convert	*into*	*multiply by*
Erg/sec	ft-lb/sec	7.3756×10^{-8}
Erg/sec	horsepower	1.341×10^{-10}
Erg/sec	kg-calorie/minute	1.433×10^{-9}
Erg/sec	metre-kgf/sec	1.02×10^{-8}
Erg/sec (*sound power*)	watt	1.0×10^{-7}
Erg/sec-cm^2 (*sound intensity*)	watt/metre^2	1.0×10^{-3}

F

To Convert	*into*	*multiply by*
^{0}F (*Fahrenheit; temperature*)	Celsius (^{0}C)	$^0C = (^0F-32)/1.8$
^{0}F (*temperature interval*)*	Kelvin (*K*)	$1K=1^0C=1.8^0F$
$^0F^{-1}$, (*1/^{0}F*)	K^{-1},(*1/K*)	1.8
^{0}F-hour-feet^2/Btu	K.m^2/W	0.176 23
Farad (*capacitance*)	abfarad	1.0×10^{-9}
Farad	cgse	8.99×10^{11}
Farad	cgsm	1.0×10^{-9}
Farad	statfarad	8.988×10^{11}
Faraday	abcoulomb	9.649×10^{3}
Faraday	ampere-hour	26.8
Faraday	coulomb	9.649×10^{4}
Faraday	statcoulomb	2.893×10^{14}
Faraday/sec	ampere (*absolute*)	9.65×10^{4}
Fathom	cable length	1/120
Fathom	chain	$0.\overline{09}$**
Fathom	feet	6.0
Fathom	in.	72.0
Fathom	metre	1.828 804
Fathom	rod	$0.\overline{36}$**
Fathom	yard	2.0
Feet	centimetre	30.48
Feet	chain	$0.0\overline{15}$**
Feet	fathom	$0.1\overline{6}$**
Feet	furlong	$0.\overline{15}$** $\times 10^{-2}$

* *For temperature interval, K may be replaced with ^{0}C if desired without changing the value or affecting the conversion factor, e.g.; kJ/(kg.K)=kJ/(kg.^{0}C)*

** *symbol — over a digit indicate recurring decimal digit*

To Convert	*into*	*multiply by*
Feet	in.	12.0
Feet (*of air*)	inch (*of water;* 4 °C)	$1.436\ 782 \times 10^{-2}$
Feet	kilometre	3.048×10^{-4}
Feet	link (*surveyor's*)	$1.5\overline{15}$*
Feet	metre	0.3048
Feet	mil	1.2×10^{4}
Feet	mile (*nautical, UK*)	1.645×10^{-4}
Feet	mile (*statute*)	1.894×10^{-4}
Feet	millimetre	3.048×10^{2}
Feet	rod	$0.0\overline{6}$*
Feet	step (*military*)	0.4
Feet	yard	$0.3\overline{3}$*
Feet²	acre	2.296×10^{-5}
Feet²	chain²	$2.295\ 684 \times 10^{-4}$
Feet²	circular mil	1.833×10^{8}
Feet²	cm²	9.29×10^{2}
Feet²	hectare	$9.293\ 68 \times 10^{-6}$
Feet²	in.²	1.44×10^{2}
Feet²	kilometre²	9.29×10^{-8}
Feet²	metre²	$9.290\ 304 \times 10^{-2}$
Feet²	mile²	3.587×10^{-8}
Feet²	millimetre²	9.29×10^{4}
Feet²	rod²	$3.673\ 09 \times 10^{-3}$
Feet²	yard²	0.1111
Feet²/hour (*viscosity*)	cm²/sec	0.258
Feet²/hour	inch²/sec	4.0×10^{-2}
Feet²/hour (*thermal diffusivity*)	metre²/hour	$9.290\ 304 \times 10^{-2}$
Feet²/hour	stoke	0.258 064

** symbol — over a digit indicate recurring decimal digit*

To Convert	*into*	*multiply by*
$Feet^2/sec$	stoke	9.2903×10^2
$Feet^3$	acre-feet	$2.295\ 68 \times 10^{-5}$
$Feet^3$	barrel (*US, liquid*)	0.178 107
$Feet^3$	bushel (*dry, UK*)	0.7786
$Feet^3$	bushel (*dry, US*)	0.803 56
$Feet^3$	cm^3	$2.831\ 684\ 6 \times 10^4$
$Feet^3$	cord	7.8125×10^{-3}
$Feet^3$	cord feet	6.25×10^{-2}
$Feet^3$	gallon (*UK*)	6.2288
$Feet^3$	gallon (*US, liquid*)	7.480 52
$Feet^3$	hogshead (*UK*)	$9.887\ 285 \times 10^{-2}$
$Feet^3$	hogshead (*US*)	0.118 738 9
$Feet^3$	$in.^3$	1.728×10^3
$Feet^3$	kilolitre	$2.831\ 61 \times 10^{-2}$
$Feet^3$	litre	28.3161
$Feet^3$	$metre^3$	$2.831\ 684\ 69 \times 10^{-2}$
$Feet^3$	perch of stone	$0.\overline{04}$*
$Feet^3$	pint (*UK*)	49.8307
$Feet^3$	pint (*US, liquid*)	59.844
$Feet^3$ (*water*)	pound	62.4
$Feet^3$	quart (*US, liquid*)	29.92
$Feet^3$	tonne (*registered*)	1.0×10^{-2}
$Feet^3$	tonne (*shipping; UK*)	$2.380\ 952 \times 10^{-2}$
$Feet^3$	tonne (*shipping; US*)	2.5×10^{-2}
$Feet^3$	$yard^3$	3.7037×10^{-2}
$Feet^3$/acre (*fodder yield*)	$metre^3$/hectare	6.99×10^{-2}
$Feet^3$/minute	cm^3/sec	$4.719\ 474 \times 102$
$Feet^3$/minute	gallon/sec	0.1247

***** *symbol* — *over a digit indicate recurring decimal digit*

To Convert	*into*	*multiply by*
Feet^3/minute	litre/sec	0.4719
Feet^3/minute	metre^3/hour	1.699
Feet^3/minute	miner's inch	0.653 594 77
Feet^3/minute	pound of water/minute	62.43
Feet^3/hour	gallon/minute	0.124 675 84
Feet^3/hour	tonne of water/ 24 hour	0.749 119 78
Feet^3/sec	gallon/minute	4.488 31 × 10^2
Feet^3/sec	litre/minute	1.698 966 9 × 10^3
Feet^3/sec	litre/sec	28.316 85
Feet^3/sec	metre^3/hour	1.019 × 10^2
Feet^3/sec	metre^3/minute	1.699 011
Feet^3/sec	millon gallon (*US*)/day	0.646 317
Feet^3/sec	pound (of water) /minute	3.745 318 4 × 10^3
Feet^3/sec	yard^3/minute	$2.22\overline{2}$*
Feet^3/pound	gallon (UK)/ pound	6.2288
Feet^3/pound	in.^{3}/pound	1.728 × 10^3
Feet^3/pound	litre/kg	62.426
Feet^3/pound	metre^3/kg	6.2428 × 10^{-2}
Feet of water	atmosphere	2.947 × 10^{-2}
Feet of water	cm of mercury	2.241 649 85
Feet of water	in. of mercury	0.8826
Feet of water	kg/cm^2	3.048 × 10^{-2}
Feet of Water	kg/metre^2	3.048 × 10^2
Feet of Water	pascal (*Pa*)	2.989 × 10^3
Feet of Water	pound/feet^2	62.43
Feet of Water	pound/in.^{2}	0.4335

* *symbol — over a digit indicate recurring decimal digit*

To Convert	*into*	*multiply by*
Feet/100 feet	percent grade	1.0
Feet/minute	cm/sec	0.508
Feet/minute	feet/sec	1.667×10^{-2}
Feet/minute	km/hour	1.829×10^{-2}
Feet/minute	metre/minute	0.3048
Feet/minute	metre/sec	5.0813×10^{-3}
Feet/minute	mile/hour	1.136×10^{-2}
Feet/hour	knot	$1.644\ 663\ 8 \times 10^{-4}$
Feet/sec	cm/sec	30.48
Feet/sec	feet/minute	60.0
Feet/sec	km/hour	1.097
Feet/sec	knot	0.5921
Feet/sec	metre/minute	18.29
Feet/sec	metre/sec	0.3048
Feet/sec	mile/hour	0.6818
Feet/sec	mile/minute	1.136×10^{-2}
Fermi (fm)	metre	1.0×10^{-15}
Feet/sec/sec	cm/sec/sec	30.48
Feet/sec/sec	km/hour/sec	1.097
Feet/sec/sec	metre/sec/sec	0.3048
Feet/sec/sec	mile/hour/sec	0.6818
Fifth (*G tone*) (*musical interval*)	cent (*natural*)	7.02×10^{2}
Fifth (*G tone*) (*musical interval*)	cent (*tempered*)	7.0×10^{2}
Fifth (*G tone*) (*musical interval*)	Savart (*natural*)	1.761×10^{2}
Fifth (*G tone*) (*musical interval*)	Savart (*tempered*)	1.756×10^{2}
Fifth (*G tone*) (*musical interval*)	Unison (*C tone*) (*natural*)	3/2

To Convert	*into*	*multiply by*
Foot candle	lumen/ft^2	1.0
Foot candle (*illuminance*)	lux (*lx*) (*lumen/metre*^{2})	10.763 91
Foot lambert (*luminance, brightness*)	cd/metre^2	3.426 259
Foot-pound	atomic mass unit	9.08×10^9
Foot-pound	Btu	1.286×10^{-3}
Foot-pound	cm-dyne	$1.355\,748\,4 \times 10^7$
Foot-pound	cm-gram	$1.382\,552\,2 \times 10^4$
Foot-pound	erg	1.356×10^7
Foot-pound	foot poundal	32.2
Foot-pound	gram-calorie	0.3238
Foot-pound	hp-hour	5.05×10^{-7}
Foot-pound	joule	1.355 817 9
Foot-pound	kg-calorie	3.24×10^{-4}
Foot-pound	kg-metre	0.1383
Foot-pound	kW-hour	3.766×10^{-7}
Foot-pound	watt-hour	3.766×10^{-4}
Foot-poundal	atomic mass unit	$2.824\,061 \times 10^8$
Foot-poundal	erg	$4.214\,075 \times 10^7$
Foot-poundal	foot-pound	$3.105\,59 \times 10^{-2}$
Foot - poundal (*energy/ mech.work/hydraulic*)	joule	$4.214\,011 \times 10^{-2}$
Foot pound/minute	Btu/minute	1.286×10^{-3}
Foot pound/minute	erg/sec	$2.259\,682\,7 \times 10^5$
Foot pound/minute	foot pound/sec	1.667×10^{-2}
Foot pound/minute	horsepower	3.03×10^{-5}
Foot pound/minute	metric horsepower	3.07×10^{-5}
Foot pound/minute	kg-calorie/minute	3.24×10^{-4}
Foot pound/minute	kilowatt	2.26×10^{-5}
Foot pound/minute	watt	2.26×10^{-2}

To Convert	*into*	*multiply by*
Foot pound/sec	Btu/hour	4.6263
Foot pound/sec	Btu/minute	7.717×10^{-2}
Foot pound/sec	erg/sec	$1.355\,822 \times 10^{7}$
Foot pound/sec	foot pound/minute	60.0
Foot pound/sec	horse power	1.818×10^{-3}
Foot pound/sec	kg-calorie/minute	1.945×10^{-2}
Foot pound/sec	kilowatt	$1.355\,818 \times 10^{-3}$
Foot pound/sec	watt	1.355 818
Foot poundf/sec	cheval vapeur	1.8434×10^{-3}
Foot poundf/sec	metre-kgf/sec	0.138 255
Fourth (*F tone*) (*musical interval*)	cent (*natural*)	4.98×10^{2}
Fourth (*F tone*) (*musical interval*)	cent (*tempered*)	5.0×10^{2}
Fourth (*F tone*) (*musical interval*)	Savart (*natural*)	1.25×10^{2}
Fourth (*F tone*) (*musical interval*)	Savart (*tempered*)	1.254×10^{2}
Fourth (*F tone*) (*musical interval*)	Unison (*C tone*) (*natural*)	4/3
Free fall (*standard*)	metre/sec²	9.807
Furlong	cable length	0.916 67
Furlong	chain	10.0
Furlong	feet	6.6×10^{2}
Furlong	kilometre	0.201 168
Furlong	link	1.0×10^{3}
Furlong	mile (*nautical, International*)	0.108 622
Furlong	mile (*nautical, UK*)	0.108 553
Furlong	mile (*US*)*	0.125
Furlong	rod	40.0
Furlong	yard	2.2×10^{2}

* *mile (US) is also statute*

G

To Convert	*into*	*multiply by*
Gallon (*UK*)	acre-ft	3.69×10^{-6}
Gallon (*US*)	acre-ft	3.07×10^{-6}
Gallon (*oil*)	barrel (*oil*)	$2.380\,952 \times 10^{-2}$
Gallon	barrel (*US; liquid*)	$3.174\,603 \times 10^{-2}$
Gallon (*US*)	cm^3	3.785×10^{3}
Gallon (*US, liquid*)	cup	16.0
Gallon (*US*)	ft^3	0.133 681
Gallon (*UK*)	ft^3	0.160 544
Gallon (*UK, Liquid*)	gallon (*US, liquid*)	1.200 95
Gallon (*US*)	gallon (*UK*)	0.832 68
Gallon (*US*)	hogshead (*US*)	$1.587\,302 \times 10^{-2}$
Gallon (*UK*)	$in.^3$	$2.774\,19 \times 10^{2}$
Gallon (*US*)	$in.^3$	$2.310\,01 \times 10^{2}$
Gallon (*UK*)	litre	4.545 96
Gallon (*US*)	litre	3.785 33
Gallon (*UK, Canadian, liquid*)	$metre^3$	4.546×10^{-3}
Gallon (*US*)	$metre^3$	$3.785\,412 \times 10^{-3}$
Gallon (*US, dry*)	$metre^3$	4.405×10^{-3}
Gallon (*US, liquid*)	ounce (*fluid*)	1.28×10^{2}
Gallon (*UK*)	pint (*UK*)	8.0
Gallon (*UK*)	pint (*US*)	9.6075
Gallon (*UK*)	pound (*lb*)	10.02
Gallon (*US*)	pint (*UK*)	6.661 4
Gallon (*US*)	pint (*US*)	8.0
Gallon (*US*)	pound (*water*)	8.3453
Gallon (*US*)	quart	4.0
Gallon (*US*)	$yard^3$	4.951×10^{-3}

To Convert	*into*	*multiply by*
Gallon(*US*)/*acre* (*application rate, pesticide*)	litre/hectare	9.353 958
Gallon (*UK*)/acre	litre/hectare	11.233 166
Gallon (*UK*)/hour	litre/hour	4.546
Gallon (*US*)/hour (*fuel consumption*)	litre/hour	3.785 412
Gallon (*UK*)/mile	litre/km	2.824 73
Gallon, million (*US*)/day	ft^3/sec	1.547
Gallon, millon (*US*)/day	gallon/minute	6.945×10^{2}
Gallon, million (*US*)/day	litre/sec	43.8
Gallon, million (*US*)/day	$metre^3$/hour	1.5773×10^{2}
Gallon/minute	ft^3/hour	8.0208
Gallon/minute	ft^3/sec	2.228×10^{-3}
Gallon/minute (*pump capacity, oil flow*)	litre/minute	3.785 412
Gallon (*US*)/minute	litre/sec	$6.309\ 02 \times 10^{-2}$
Gallon (*UK*)/minute	litre/sec	$7.57\overline{6}$* $\times 10^{-2}$
Gallon (*US*)/minute	$metre^3$/hour	0.227 117 87
Gallon (*US, liquid*)/minute	$metre^3$/sec	6.309×10^{-5}
Gallon(*US*)/minute	million gallon (*US*)/day	1.44×10^{-3}
Gallon/minute	tonne of water/ 24 hour	6.008 532 12
Gallon (*UK*)/pound	ft^3/pound	0.160 544
Gallon (*UK*)/pound	$in.^3$/pound	$2.774\ 19 \times 10^{2}$
Gallon (*UK*)/pound	litre/kg	10.0221
Gallon (*UK*)/pound	$metre^3$/kg	$1.002\ 24 \times 10^{-2}$
Gallon/sec	ft^3/minute	8.019 246 19
Gallon/sec	litre/minute	$2.271\ 178\ 7 \times 10^{2}$
Gallon (*US*)/sec	$yard^3$/metre	0.297 000 30

***** *symbol — over a digit indicate recurring decimal digit*

To Convert	*into*	*multiply by*
Gauss	cgse*	$3.3\overline{3}$** × 10^{-11}
Gauss	line/cm^2	1.0
Gauss	line/$in.^2$	6.452
Gauss	tesla	1.0 × 10^{-4}
Gauss	weber/cm^2	1.0 × 10^{-8}
Gauss	weber/$in.^2$	6.452 × 10^{-8}
Gauss (*magnetic flux density or magnetic induction*)	weber/$metre^2$ (*tesla*)	1.0 × 10^{-4}
Gilbert (*magnetomotive force*)	abampere-turn	7.958 × 10^{-2}
Gilbert	ampere-turn	0.7958
Gilbert	cgse	3.0 × 10^{10}
Gilbert/cm	ampere-turn/cm	0.7958
Gilbert/cm	ampere-turn/in.	2.021
Gilbert/cm	ampere-turn/m	79.58
Gill*** (*UK*)	centilitre	11.834 319 53
Gill (*UK*)	cm^3	1.420 61 × 10^2
Gill	$in.^3$	8.669 342 6
Gill	litre	0.142 065
Gill	millilitre	1.420 65 × 10^2
Gill	minim (*US*)	2.305 81 × 10^3
Gill	minim (*UK*)	2.4 × 10^3
Gill	ounce (*UK, fluid*)	5.0
Gill	ounce (*US, liquid*)	4.803 77
Gill	pint (*liquid*)	0.25
Gon	degree (*angle*)	0.900 900 9
Gon	radian	1.569 859 × 10^{-2}
Grade	radian	1.571 × 10^{-2}

* *cgse : electrostatic system also known as electrostatic unit (esu)*

** *symbol — over a digit indicate recurring decimal digit*

*** *Gill is a unit of liquid measure, equal to a quarter of a pint*

To Convert	*into*	*multiply by*
Gram	carat (*metric*)	5.0
Gram	centigram	1.0×10^{2}
Gram	dalton	$6.060\,606 \times 10^{23}$
Gram	decigram	10.0
Gram	dekagram	0.1
Gram	drachm (*apothecarie's*)	0.257 206
Gram	dram (*avoirdupois*)	0.564 38
Gram	dyne	$9.806\,65 \times 10^{2}$
Gram	grain	15.4324
Gram	hectogram	1.0×10^{-2}
Gram	kilogram	1.0×10^{-3}
Gram	milligram	1.0×10^{3}
Gram	ounce (*avoirdupois*)	3.5274×10^{-2}
Gram	ounce (*apothe-carie's or troy*)	$3.215\,07 \times 10^{-2}$
Gram	penny weight (*troy*)	0.643 016 52
Gram	pound	2.205×10^{-3}
Gram	pound (*troy*)	$2.679\,23 \times 10^{-3}$
Gram	poundal	7.093×10^{-2}
Gram	tola	8.5735×10^{-2}
Gram - calorie	British thermal unit	3.9683×10^{-3}
Gram - calorie	erg	4.1868×10^{7}
Gram - calorie	ft-pound	3.088
Gram - calorie	horsepower-hour	1.5596×10^{-6}
Gram - calorie	joule	4.1868
Gram - calorie	kilowatt-hour	1.163×10^{-6}
Gram - calorie	watt-hour	$1.162\,99 \times 10^{-3}$
Gram - calorie/sec	Btu/hour	14.286
Gram - cm	Btu	9.297×10^{-8}

To Convert	*into*	*multiply by*
Gram - cm	erg	$9.806\,65 \times 10^{2}$
Gram - cm	joule	$9.806\,65 \times 10^{-5}$
Gram - cm	kg-calorie	2.343×10^{-8}
Gram - cm	kg-metre	1.0×10^{-5}
Gram/cm	pound/in.	5.6×10^{-3}
Gram/cm^2	pound/ft^2	2.0481
Gram/cm^3	gram/millilitre	1.000 053
Gram/cm^3	kilogram/m^3	1.0×10^{3}
Gram/cm^3	pound/ft^3	62.43
Gram/cm^3	pound/gallon (*density*)	$1.002\,245 \times 10^{-2}$
Gram/cm^3	pound/in.^{3}	3.613×10^{-2}
Gram/cm^3	pound/mil-ft	3.405×10^{-7}
Gram/cm-sec	poise (*absolute viscosity*)	1.0
Gram force	dyne	$9.806\,65 \times 10^{2}$
Gram force	joule/cm	$9.806\,65 \times 10^{-5}$
Gram force	joule/metre (*newton*)	$9.806\,65 \times 10^{-3}$
Gram force-sec^2 (*Newtons' law conversion factor*)	gram mass-cm	$9.806\,65 \times 10^{2}$
Gram/kWh	pound/(*hp-hour*)	$1.643\,99 \times 10^{-3}$
Gram/litre	grain/gallon	58.417
Gram/litre	part/million (*ppm*)	1.0×10^{3}
Gram/litre	pound/ft^3	6.2427×10^{-2}
Gram/litre	pound/gallon (*UK*)	$1.002\,215 \times 10^{-2}$
Gram/litre	pound/thousand gallon	8.345
Gram mass-cm	gram force - sec^2	$1.019\,72 \times 10^{-3}$
Gram-metre	ounce-in.	1.388 738 83

To Convert	*into*	*multiply by*
Gram/metre^2	ounce/ft^2	$3.277\ 06 \times 10^{-3}$
Gram/metre^2	ounce/yard^2	$2.949\ 352 \times 10^{-2}$
Gram/metre^3	grain/ft^3	0.436 996
Gram/metre^3	grain/gallon (*UK*)	7.0157×10^{-2}
Gram/metre^3	milligram/litre	1.000 028
Gram/metre^3	ounce (*avoirdupois*)/ft^3	$9.988\ 513\ 2 \times 10^{-4}$
Gram/metre^3	ounce (*avoirdupois*)/ gallon (*UK*)	1.6036×10^{-4}
Gram/millilitre	gram/cm^3	0.999 947
Gram/millilitre	pound/in.^{3}	$3.612\ 625 \times 10^{-2}$
Gram/MJ	pound/(*hp-hour*)	$5.918\ 35 \times 10^{-3}$
Grain	carat	0.323 994 73
Grain	drachm	$1.6\overline{6}$* $\times 10^{-2}$
Grain	dram (*avoirdupois*)	$3.657\ 143 \times 10^{-2}$
Grain (*troy*)	grain (*avoirdupois*)	1.0
Grain (*troy*)	gram	6.4799×10^{-2}
Grain	milligram	64.799
Grain (*troy*)	ounce (*avoirdupois*)	2.0833×10^{-3}
Grain (*troy*)	pennyweight (*troy*)	4.167×10^{-2}
Grain	pound	1/7000
Grain	pound (*troy*)	1.7361×10^{-4}
Grain	scruple	5.0×10^{-2}
Grain	tola	$5.\overline{5}$* $\times 10^{-3}$
Grain/ft^3	grain/gallon (*UK*)	0.160 544
Grain/ft^3	gram/metre^3	2.288 35
Grain/ft^3	milligram/litre	2.288 42
Grain/ft^3	ounce (*avoirdupois*)/ft^3	$2.285\ 71 \times 10^{-3}$

**** symbol — over a digit indicate recurring decimal digit***

To Convert	*into*	*multiply by*
Grain/ft³	ounce (*avoirdupois*)/ gallon (*UK*)	$3.669\ 57 \times 10^{-4}$
Grain/gallon (*UK*)	grain/ft³	6.2288
Grain/gallon	gram/litre	1.7118×10^{-2}
Grain/gallon (*UK*)	gram/metre³	14.2538
Grain/gallon (*UK*)	milligram/litre	14.2542
Grain/gallon (*US*)	milligram/litre	17.118
Grain/gallon (*UK*)	ounce (*avoirdupois*)/ft³	$1.423\ 73 \times 10^{-2}$
Grain/gallon (*UK*)	ounce (*avoirdupois*)/ gallon(*UK*)	$2.285\ 71 \times 10^{-3}$
Grain/gallon (*UK*)	parts/million (*ppm*)	14.286
Grain/gallon (*US*)	parts/million (*ppm*)	17.118
Grain/gallon (*US*)	pound/million gallon	1.4286×10^{2}
Gray (*J/kg*)	rad (*absorbed dose*)	1.0×10^{2}

H

To Convert	*into*	*multiply by*
Hand	cm	10.16
Hand	in.	4.0
Hardness number*		
Hectare (*ha*)	acre	2.471 05
Hectare	ft^2	1.076×10^5
Hectare	$metre^2$	1.0×10^4
Hectare	$mile^2$	$3.861\ 02 \times 10^{-3}$
Hectare	rod^2	$3.953\ 69 \times 10^2$
Hectare	rood	9.8842
Hectare	$yard^2$	$1.195\ 99 \times 10^4$
Hectogram	gram	1.0×10^2
Hectolitre	litre	1.0×10^2
Hectometre	metre	1.0×10^2
Hectowatt	watt	1.0×10^2
Hemisphere	sphere	0.5
Hemisphere	spherical right angle	4.0
Hemisphere	steradian	6.283
Henrie (*inductance*)	abhenrie	1.0×10^9
Henrie	cgse	1.11×10^{-12}
Henrie	cgsm	1.0×10^9
Henrie	millihenrie	1.0×10^3
Henrie	stathenrie	1.112×10^{-12}
Hertz	cycle/sec	1.0
Hogshead	barrel	2.0
Hogshead (*UK*)	ft^3	10.114
Hogshead (*US*)	ft^3	8.421 84
Hogshead (*US*)	gallon (*US*)	63.0

* *conventional hardness numbers, BHN, R etc. are not affected by change to SI*

To Convert	*into*	*multiply by*
Hour (*h*)	arc (*angular; earth*)	37.5
Hour	day (*d*)	4.167×10^{-2}
Hour	minute	60.0
Hour	second	3.6×10^{3}
Hour	week	5.952×10^{-3}
Horsepower (*hp; boiler*)	Btu/hour	3.3479×10^{4}
Horsepower	Btu/min	42.44
Horsepower	cheval vapeur	1.013 87
Horsepower	erg/sec	$7.457\,121\,6 \times 10^{9}$
Horsepower	ft-lb/min	3.3×10^{4}
Horsepower	ft-lb/sec	5.5×10^{2}
Horsepower (550 *ft-lb/sec*)	horsepower (*metric; 542.5 ft-lb/sec*)	1.014
Horsepower	in.-lb/sec	6.6×10^{3}
Horsepower	kg-calorie/min	10.68
Horsepower	kilowatt	$0.745\,6\overline{9}$*
Horsepower	metre-kgf/sec	76.04
Horsepower	watt	$7.456\overline{9}$* $\times 10^{2}$
Horsepower (*boiler*)	kilowatt	9.803
Horsepower (*metric*)**	Btu/min	41.8
Horsepower (*metric*)	ft-lb/min	3.255×10^{4}
Horsepower (*metric; 542.5 ft-lb/sec*)	horsepower (*550 ft-lb/sec*)	0.9863
Horsepower (*metric*)	kilowatt	0.736
Horsepower (*metric*)	watt	7.36×10^{2}
Horsepower-hour	Btu	$2.544\,43 \times 10^{3}$
Horsepower-hour	erg	2.6845×10^{13}
Horsepower-hour	ft-lb	1.98×10^{6}

* *symbol — over a digit indicate recurring decimal digit*

** *also called as Pferde Starke (PS)*

To Convert	*into*	*multiply by*
Horsepower-hour	gram-calorie	6.4119×10^5
Horsepower-hour	joule	$2.684\ 52 \times 10^6$
Horsepower-hour	kg-calorie	6.4119×10^2
Horsepower-hour	kgf-metre	2.737×10^5
Horsepower-hour	killowatt-hour	$0.745\ 69\overline{9}$*
Horsepower-hour (*energy, mechanical*)	megajoule	2.684 52
Horsepower-hour	watt-hour	$7.456\overline{9}$* $\times 10^2$
Horsepower-hour/gallon (*fuel economy*)	killowatt-hour/ litre	0.196 993 1
Hundred weight (*cwt*)	mound	1.361 118 29
Hundred weight (*long; cwt*)	pound	1.12×10^2
Hundred weight (*long*)	tonne (*long*)	5.0×10^{-2}
Hundred weight (*short*)	ounce (*avoirdupois*)	1.6×10^3
Hundred weight (*short*)	pound	1.0×10^2
Hundred weight (*short*)	tonne (*long*)	$4.464\ 29 \times 10^{-2}$
Hundred weight (*short*)	tonne (*metric*)	$4.535\ 92 \times 10^{-2}$

* *symbol — over a digit indicate recurring decimal digit*

I

To Convert	*into*	*multiply by*
Inch	angstrom unit (Å)	$2.540\,005 \times 10^{8}$
Inch	cm	2.54
Inch	chain	$1.\overline{26}^{*} \times 10^{-3}$
Inch	cubit	$5.\overline{5}^{*} \times 10^{-2}$
Inch	cubit (*bible*)	$4.587\,156 \times 10^{-2}$
Inch	ell	$2.\overline{2}^{*} \times 10^{-2}$
Inch	em, pica	6.0
Inch	fathom	$1.3\overline{8}^{*} \times 10^{-2}$
Inch	ft	$8.3\overline{3}^{*} \times 10^{-2}$
Inch	hand	0.25
Inch	kilometre	2.54×10^{-5}
Inch	link (*Engineer's*)	$8.3\overline{3}^{*} \times 10^{-2}$
Inch	link (*Surveyor's*)	$0.126\,\overline{26}^{*}$
Inch	metre	2.54×10^{-2}
Inch	mile	1.578×10^{-5}
Inch	mile (*nautical*)	$1.371\,49 \times 10^{-5}$
Inch	millimetre	25.4
Inch	mil	1.0×10^{3}
Inch	nail	$0.4\overline{4}^{*}$
Inch	palm	$0.3\overline{3}^{*}$
Inch	rod	$5.\overline{05}^{*} \times 10^{-3}$
Inch	span	$0.1\overline{1}^{*}$
Inch	yard	$2.7\overline{7}^{*} \times 10^{-2}$
Inch (*of mercury*)	atmosphere	$3.342\,11 \times 10^{-2}$
Inch (*of mercury*)	dyne/cm^2	$.386\,386\,7 \times 10^{4}$
Inch (*of mercury*)	feet (*of water*)	1.133
Inch (*of mercury*)	inch (*of water; 4°C*)	13.595 084

* *symbol — over a digit indicate recurring decimal digit*

To Convert	*into*	*multiply by*
Inch (*of mercury*)	kg/cm²	$3.453\ 16 \times 10^{-2}$
Inch (*of mercury*)	kg/m²	$3.453\ 16 \times 10^{2}$
Inch (*of mercury*)	metre (*of water*)	0.345 616
Inch (*of mercury*)	millibar	33.8639
Inch (*of mercury*)	pound/ft²	70.726
Inch (*of mercury*)	pound/in.²	0.4912
Inch (*of mercury; 32°F*)	pascal	$3.386\ 389 \times 10^{3}$
Inch (*of mercury; 60°F*)	pascal	$3.376\ 85 \times 10^{3}$
Inch (*of water; 4°C*)	atmosphere	$2.458\ 32 \times 10^{-3}$
Inch (*of water; 4°C*)	dyne/cm²	$2.490\ 66 \times 10^{3}$
Inch (*of water; 4°C*)	feet (*of air*)	69.6
Inch (*of water; 4°C*)	in. (*of mercury*)	7.3556×10^{-2}
Inch (*of water; 4°C*)	kg/cm²	2.54×10^{-3}
Inch (*of water; 4°C*)	millibar	2.490 89
Inch (*of water; 4°C*)	mm of mercury	1.868 320 75
Inch (*of water; 4°C*)	ounce/in.²	0.5781
Inch (*of water; 4°C*)	pascal	$2.490\ 82 \times 10^{-2}$
Inch (*of water; 4°C*)	pound/ft²	5.2023
Inch (*of water; 4°C*)	pound/in.²	3.613×10^{-2}
Inch (*of water; 60°F*)	pascal	2.4884×10^{-2}
Inch²	acre	$1.594\ 225 \times 10^{-7}$
Inch²	cm²	6.4516
Inch²	cicular mil	1.273×10^{6}
Inch²	ft²	6.944×10^{-3}
Inch²	kilometre²	6.4516×10^{-10}
Inch²	metre²	6.4516×10^{-4}
Inch²	mil²	1.0×10^{6}
Inch²	mile²	$2.490\ 66 \times 10^{-10}$
Inch²	mm²	6.4516×10^{2}
Inch²	yard²	7.716×10^{-4}
Inch²/sec	ft²/hour	25.0

To Convert	*into*	*multiply by*
$Inch^3$	barrel (*US; dry*)	$1.417\,233\,6 \times 10^{-4}$
$Inch^3$	bushel	4.651×10^{-4}
$Inch^3$	centilitre	1.638 538 4
$Inch^3$ (*section modulus*)	cm^3	16.387 06
$Inch^3$	ft^3	5.787×10^{-4}
$Inch^3$	gallon (*UK*)	$3.604\,65 \times 10^{-3}$
$Inch^3$	gallon (*US*)	$4.328\,98 \times 10^{-3}$
$Inch^3$	gill (*UK*)	0.115 349
$Inch^3$	litre	$1.638\,66 \times 10^{-2}$
$Inch^3$	$metre^3$	$1.638\,706 \times 10^{-5}$
$Inch^3$	mil-ft	1.061×10^{5}
$Inch^3$	millilitre	16.385 384
$Inch^3$	minim (*UK*)	$2.768\,37 \times 10^{2}$
$Inch^3$	minim (*US*)	$2.659\,72 \times 10^{2}$
$Inch^3$	ounce (*fluid; UK*)	0.576 74
$Inch^3$	ounce (*fluid; US*)	0.554 11
$Inch^3$	peck (*UK*)	1.8031×10^{-3}
$Inch^3$	peck (*US*)	1.8601×10^{-3}
$Inch^3$	pint (*dry*)	$2.976\,19 \times 10^{-2}$
$Inch^3$	pint (*UK*)	$2.883\,72 \times 10^{-2}$
$Inch^3$	pint (*US*)	$3.463\,18 \times 10^{-2}$
$Inch^3$	pound of water	3.61×10^{-2}
$Inch^3$	quart (*liquid*)	$1.732\,18 \times 10^{-2}$
$Inch^3$	quart (*dry*)	$1.488\,095 \times 10^{-2}$
$Inch^3$	$yard^3$	2.143×10^{-5}
Inch-pound	Btu	$1.070\,764\,7 \times 10^{-4}$
Inch-pound	calorie	$2.695\,418 \times 10^{-2}$
Inch-pound	kilowatt hour	$3.139\,717\,4 \times 10^{-8}$
Inch-pound force	joule	0.112 994 35
Inch-lb/in.^{3}-^{0}F	Btu/in.^{3}-^{0}F (*volumetric*	$1.071\,811\,4 \times 10^{-4}$

To Convert	*into*	*multiply by*
	specific heat)	
Inch-lb/in.^{2}-s-^{0}F/in.	Btu-ft^2-h-^{0}F/ft (*thermal conductance*)	4.629 629 63
Inch-lb/sec	horsepower	$1.5\overline{1}$* × 10^{-4}
Inch-lb/sec	kilowatt	1.129 943 5 × 10^{-4}
Inch^3/pound	ft^3/pound	0.5787 × 10^{-3}
Inch^3/pound	gallon (*UK*)/pound	3.604 65 × 10^{-3}
Inch^3/pound	litre/kg	3.612 63 × 10^{-2}
Inch^3/pound	metre^3/kg	3.612 73 × 10^{-5}
Inch4 (*second moment of area*)	cm^4	41.623 14
Inch^4	mm^4	4.162 314 × 10^5
International ampere	ampere (*absolute*)	0.9998
International volt	joule	9.654 × 10^4
International volt	joule (*absolute*)	1.593 × 10^{-19}
International volt	volt (*absolute*)	1.0003

* *symbol* — *over a digit indicate recurring decimal digit*

J

To Convert	*into*	*multiply by*
Joule	atomic mass unit	6.702×10^{9}
Joule	Btu	9.4782×10^{-4}
Joule	Btu (*mean*)	$9.470\,862\,89 \times 10^{-4}$
Joule	calorie (*mean*)	0.238 662 35
Joule	calorie, gram (*thermo-chemical*)	0.238 845 9
Joule	electron volt	$6.241\,457 \times 10^{18}$
Joule	erg	1.0×10^{7}
Joule	ft-lbf	0.7376
Joule	ft-poundal	23.73
Joule	gram-cm	$1.019\,716\,2 \times 10^{4}$
Joule	horsepower-hour	3.78×10^{-7}
Joule	in.-lbf	8.85
Joule	kg-calorie	$2.388\,46 \times 10^{-4}$
Joule	kWh	$2.7\overline{7}^{*} \times 10^{-7}$
Joule	metre-kgf	0.102
Joule	volt (*International*)	$1.035\,840\,07 \times 10^{-5}$
Joule (*absolute*)	volt (*International*)	$6.277\,463\,9 \times 10^{18}$
Joule	watt-hour	2.778×10^{-4}
Joule	watt-sec	1.0
Joule/cm	dyne	1.0×10^{7}
Joule/cm	gram-f	1.02×10^{4}
Joule/cm	joule/metre (*newton*)	1.0×10^{2}
Joule/cm	kilogram	10.197 162 13
Joule/cm	pound-force	22.48
Joule/cm	poundal	7.23×10^{2}
Joule/cm^{3}-°C	Btu/ft^{3}-°F	14.9107

** symbol — over a digit indicate recurring decimal digit*

To Convert	*into*	*multiply by*
Joule/cm³-°C (*specific heat, volume basis*)	kcalorie/m³-°C	$2.388\ 46 \times 10^{2}$
Joule/gram	Btu/lb	0.429 923
Joule/gram (*enthalpy*)	kilocalorie/ kilogram	0.238 846
Joule/gram-°C	Btu/lb-°F	0.238 846
Joule/gram-°C (*specific heat; mass basis*)	calorie/gram-°C	0.238 846
Joule/kg	calorie/gram (*specific heat*)	$2.388\ 458\ 97 \times 10^{-4}$
Joule/metre (*newton*)	dyne	1.0×10^{5}
Joule/metre (*newton*)	gram force	$1.019\ 716\ 2 \times 10^{2}$
Joule/metre (*newton*)	joule/cm	1.0×10^{-2}
Joule/metre (newton)	kilogram	0.101 971 62
Joule/metre (*newton*)	pound (*avoirdupois*)	0.224 820 14
Joule/metre (*newton*)	poundal	7.233 011 46
Joule/m²	Btu/ft² (*solar radiant exposure*)	$8.805\ 508\ 16 \times 10^{-5}$
Joule/m²	langley	$2.390\ 057\ 36 \times 10^{-5}$
Joule/m³	erg/cm³ (*sound energy density*)	10.0
Joule/sec	watt (*absolute*)	1.0

K

To Convert	*into*	*multiply by*
Kelvin (*temperature; K*)	celsius (°C)	°C=(K-273.15)
Kelvin (*K*)	fahrenheit (°F)	°F=(1.8 K-459.67)
Kelvin (*K*)	rankin (*R*)	1.8
Kelvin-m²/W	°F-hour-ft²/Btu	5.674 402 77
Kilocalorie (heat)	kilojoule	4.1868
Kilocalorie/hour	Btu/hour	3.968 32
Kilocalorie/hour	calorie/sec	0.277$\overline{7}$*
Kilocalorie/hour	watt	1.163
Kilocalorie/kg	Btu/pound	1.8
Kilocalorie/kg	joule/gram	4.1868
Kilocalorie/metre-hour-°C	Btu/ft-hour-°F	0.671 97
Kilocalorie/metre-hour-°C	Btu-in./ft²-hour-°F	8.0636
Kilocalorie/metre-hour-°C	calorie/cm-sec-°C	2.77$\overline{7}$* $\times 10^{-3}$
Kilocalorie/metre-hour-°C	watt/cm-°C	11.63×10^{-3}
Kilocalorie/metre²-hour	Btu/ft²-hour	0.368 669
Kilocalorie/metre²-hour	calorie/cm²-sec	2.7$\overline{7}$* $\times 10^{-5}$
Kilocalorie/metre²-hour	watt/cm²	1.163×10^{-4}
Kilocalorie/metre²-hour-°C (*thermal conductance*)	Btu/ft²-hour-°F	0.204 816
Kilocalorie/metre²-hour-°C	calorie/cm²-sec-°C	2.7$\overline{7}$* $\times 10^{-5}$
Kilocalorie/metre²-hour-°C	watt/cm²-°C	1.163×10^{-4}
Kilocalorie/metre³-°C (*specific heat, volume basis*)	Btu/ft³-°F	6.2428×10^{-2}
Kilocalorie/metre³-°C	joule/cm³-°C	4.1868×10^{-3}
Kilogauss (*density; magnetic flux*)	tesla (T)	0.1
Kilogram (kg)	chemical mass unit	$6.023\,225\,6 \times 10^{26}$
Kilogram	cwt	$1.968\,41 \times 10^{-2}$

***** *symbol — over a digit indicate recurring decimal digit*

To Convert	*into*	*multiply by*
Kilogram	gram	1.0×10^3
Kilogram	joule/cm	$9.806\ 65 \times 10^{-2}$
Kilogram	joule/metre (*newton*)	9.806 65
Kilogram	maund	$2.679\ 23 \times 10^{-2}$
Kilogram	metre of water	0.101 971 62
Kilogram (mass)	metric technical weight	0.101 972
Kilogram	millilitre (water)	$1.000\ 03 \times 10^{-3}$
Kilogram	myria gram	0.1
Kilogram	pound	2.204 622 622
Kilogram	quintal (*q*)	1.0×10^{-2}
Kilogram	seer	1.071 69
Kilogram (mass)	slug	6.8522×10^{-2}
Kilogram	tonne (*long*)	9.842×10^{-4}
Kilogram	tonne (*metric*)	1.0×10^{-3}
Kilogram	tonne (*short*)	$1.102\ 31 \times 10^{-3}$
Kilogram-calorie	Btu	3.968
Kilogram-calorie	erg	$4.185\ 851\ 8 \times 10^{10}$
Kilogram-calorie	foot-pound	3.088×10^3
Kilogram-calorie	gram-cm	$4.268\ 032\ 4 \times 10^7$
Kilogram-calorie	horsepower-hour	1.56×10^{-3}
Kilogram-calorie	joule	4.186×10^3
Kilogram-calorie	kilogram-metre	4.269×10^2
Kilogram-calorie	kilowatt-hour	1.163×10^{-3}
Kilogram-calorie	watt-hour	1.162 115
Kilogram-calorie/minute	erg/sec	$6.978\ 367 \times 10^8$
Kilogram-calorie/minute	ft-pound/minute	$3.086\ 419\ 8 \times 10^3$
Kilogram-calorie/minute	ft-pound/sec	51.413 881 75
Kilogram-calorie/minute	horse power	$9.363\ 296 \times 10^{-2}$
Kilogram-calorie/minute	kilowatt	6.9735×10^{-2}

To Convert	*into*	*multiply by*
Kilogram/cm^2	atmosphere	0.9678
Kilogram/cm^2	dyne/cm^2	$9.806\,65 \times 10^5$
Kilogram/cm^2	ft of water	32.81
Kilogram/cm^2	in. of mercury	28.96
Kilogram/cm^2	in. of water (4^oC)	$3.937\,007\,87 \times 10^2$
Kilogramf/cm^2	kilo-pascal (kPa)	98.0665
Kilogram/cm^2	metre of water	10.0
Kilogramf/cm^2	pieze	98.0665
Kilogram/cm^2	pound/in.^{2}	14.22
Kilogram/cm^2	pound/ft^2	$2.048\,16 \times 10^3$
Kilogram/cm^2	tonne/in.^{2}	6.3497×10^{-3}
Kilogram-force (kgf)	dyne	$9.806\,65 \times 10^5$
Kilogram-force	megadyne	0.980 665
Kilogram-force (kgf)	newton	9.806 65
Kilogram-force	poundal	70.932
Kilogram force-cm (*moment of force*)	newton-metre	$9.806\,65 \times 10^{-2}$
Kilogram/hectare (*crop yield*)	bushel (60 lb)/acre	$1.486\,768 \times 10^{-2}$
Kilogram/hectare (*crop yield*)	pound/acre	0.89
Kilogram/hectare	tonne/acre	$3.982\,477 \times 10^{-4}$
Kilogram/litre	pound/gallon	8.345 406 35
Kilogram-metre	Btu	9.297×10^{-3}
Kilogram-metre	erg	9.807×10^7
Kilogram-metre	foot-pound	7.233
Kilogram-metre	gram-cm	1.0×10^5
Kilogram-metre	horsepower-hour	3.653×10^{-6}
Kilogram-metre	joule	9.807
Kilogram-metre	kilogram-calorie	2.343×10^{-3}
Kilogram-metre	kilowatt-hour	2.723×10^{-6}

To Convert	*into*	*multiply by*
Kilogram-metre	watt-hour	2.723×10^{-3}
Kilogram-metre/sec	pound-ft/sec	7.233 011 46
Kilogram-metre^2	pound-ft^2	23.730 366
Kilogram-metre^2/sec	pound-ft^2/sec	23.730 366
Kilogram/metre	pound/ft	0.671 97
Kilogram/metre	pound/yard	2.015 906 71
Kilogram/metre^2	atmosphere	9.6784×10^{-5}
Kilogram/metre^2	bar	98.0665×10^{-6}
Kilogram/metre^2	cm of mercury	$7.352\ 94 \times 10^{-3}$
Kilogram/metre^2	ft of water	3.281×10^{-3}
Kilogram/metre^2	in. of mercury	2.8959×10^{-3}
Kilogram/metre^2	pound/ft^2	0.2048
Kilogram/metre^2	pound/in.^{2}	1.422×10^{-3}
Kilogram/metre^2	tonne (short)/ft^2	$1.024\ 065\ 54 \times 10^{-4}$
Kilogram/metre^3	gram/cm^3	1.0×10^{-3}
Kilogram/metre^3	pound/ft^3	6.243×10^{-2}
Kilogram/metre^3	pound/in.^{3}	3.613×10^{-5}
Kilogram/metre^3	pound/mil-ft	3.405×10^{-10}
Kilogram/metre^3	pound/yard^3	1.685 555 28
Kilogram/metre^3	slug/ft^3	$1.940\ 32 \times 10^{-3}$
Kilogram/millimetre^2	kilogram/metre^2	1.0×10^{6}
Kilogram/pascal-sec-metre^2	perm (0°C)	$1.747\ 946\ 2 \times 10^{10}$
Kilogram/second	tonne(UK)/hour	3.543 14
Kilogram/second	tonne(metric)/hour	3.6
Kilogramf-metre	calorie	2.341 920 37
Kilogramf/metre	millibar	$9.806\ 614 \times 10^{-2}$
Kilogramf/metre^2	mm of mercury	$7.355\ 591 \times 10^{-2}$
Kilogramf/metre^2	pascal (Pa)	9.806 65
Kilogramf-sec/metre^2	poise	98.0665
Kilogramf-sec/metre^2	poundal-sec/ft^2	6.5898

To Convert	*into*	*multiply by*
Kilogramf-sec/metre^2	poundf-sec/ft^2	0.204 816
Kilojoule	kilocalorie	0.238 845 9
Kilojoule/K	Btu/°F	0.526 564 94
Kilojoule/kg	Btu/lb	0.429 922 61
Kilojoule/litre	Btu/gallon (*calorific value*)	4.308 487 72
Kilojoule/m^3	Btu/ft^3	$2.683\ 843 \times 10^{-2}$
Kilolines (*magnetic flux*)	maxwell (*lines*)	1.0×10^3
Kilolines	weber (*Wb*)	1.0×10^{-5}
Kilolitre	acre-ft	8.1074×10^{-4}
Kilolitre	barrel (*US; liquid*)	6.2899
Kilolitre	bushel (*UK*)	27.4969
Kilolitre	bushel (*US; dry*)	28.3782
Kilolitre	ft^3	35.3157
Kilolitre	metre^3	1.000 028
Kilolitre	yard^3	1.307 99
Kilometre	astronomical unit	$6.688\ 963\ 2 \times 10^{-9}$
Kilometre	cable length (*UK*)	4.556 72'
Kilometre	centimetre	1.0×10^5
Kilometre	ft	3.281×10^3
Kilometre	furlong	4.970 97
Kilometre	in.	3.937×10^4
Kilometre	light year	$1.056\ 980\ 8 \times 10^{-13}$
Kilometre	metre	1.0×10^3
Kilometre	mil	3.937×10^7
Kilometre	mile	0.621 371 192 2
Kilometre	mile (*nautical; International*)	0.539 96
Kilometre	mile (*nautical; UK*)	0.539 61
Kilometre	millimetre	1.0×10^6
Kilometre	myria metre	0.1

To Convert	*into*	*multiply by*
Kilometre	parsec	$3.242\,542\,2 \times 10^{-14}$
Kilometre	yard	$1.093\,61 \times 10^{3}$
Kilometre^2	acre	2.471×10^{2}
Kilometre^2	cm^2	1.0×10^{10}
Kilometre^2	ft^2	10.76×10^{6}
Kilometre^2	in.^{2}	1.55×10^{9}
Kilometre^2	metre^2	1.0×10^{6}
Kilometre^2	mile^2	0.3861
Kilometre^2	mm^2	1.0×10^{12}
Kilometre^2	yard^2	1.196×10^{6}
Kilometre/hour	cm/sec	27.78
Kilometre/hour	ft/minute	54.68
Kilometre/hour	ft/sec	0.9113
Kilometre/hour	knot	0.5396
Kilometre/hour	metre/minute	16.67
Kilometre/hour	metre/sec	$0.27\overline{7}$*
Kilometre/hour	mile/hour	0.621 37
Kilometre/hour/sec	cm/sec/sec	27.78
Kilometre/hour/sec	ft/sec/sec	0.9113
Kilometre/hour/sec	metre/sec/sec	0.2778
Kilometre/hour/sec	mile/hour/sec	0.6214
Kilometre/litre (*fuel consumption*)	mile/gallon (*UK*)	2.824 73
Kilometre/litre	mile/gallon (*US*)	2.408 804 47
Kilometre/minute	metre/sec	$16.6\overline{6}$*
Kilometre/minute	mile/hour	37.285 607 76
Kilometre/minute	mile/minute	0.621 371 19
Kilonewton (*kN*)	tonne force	0.100 361 3
Kilopascal	kilogramf/cm^2	$1.019\,716 \times 10^{-2}$
Kilopascal	pound/ft^2	20.885 459

** symbol — over a digit indicate recurring decimal digit*

To Convert	*into*	*multiply by*
Kilopascal	pound/in.^{2}	0.145 037 74
Kilowatt (kW)	Btu/minute	56.92
Kilowatt	cheval vapeur*	1.359 62
Kilowatt	ft-pound/minute	4.426×10^4
Kilowatt	ft-pound/sec	7.3756×10^2
Kilowatt	horsepower	1.341 02
Kliowatt	horsepower (*boiler*)	0.102 009 59
Kilowatt	horsepower (*metric*)	1.36
Kilowatt	in.-pound/sec	8.85×10^3
Kilowatt	kg-calorie/minute	14.34
Kilowatt	metre-kgf/second	$1.019\ 72 \times 10^2$
Kilowatt	myria watt	0.1
Kilowatt	ton (refrigeration)	0.284 333 24
Kilowatt	watt	1.0×10^3
Kilowatt-hour	atomic mass unit	2.412×10^{16}
Kilowatt-hour	Btu	$3.412\ 14 \times 10^3$
Kilowatt-hour	calorie	$8.598\ 452\ 3 \times 10^5$
Kilowatt-hour	erg	3.6×10^{13}
Kilowatt-hour	ft-pound	$2.654\ 16 \times 10^6$
Kilowatt-hour	gram-calorie	8.5985×10^5
Kilowatt-hour	horsepower-hour	1.341 02
Kilowatt-hour	in.-pound	3.185×10^7
Kilowatt-hour	joule (J)	3.6×10^6
Kilowatt-hour	kg-calorie	8.5985×10^2
Kilowatt-hour	kg-metre	3.671×10^5
Kilowatt-hour	megajoule (MJ)	3.6
Kilowatt-hour	pound of water (*evaporated from and at 212°F*)	3.53

* *This unit is also called as "metric horsepower" or "Pferdekraft"*

To Convert	*into*	*multiply by*
Kilowatt-hour	pound of water (*raised from 62°* *to 212°F*)	22.75
Kilowatt-hour/litre	horsepower-hour/ gallon (*fuel economy*)	5.076 319 93
Knot	cm/sec	51.466 803 91
Knot	ft/hour	$6.080\ 27 \times 10^3$*
Knot	ft/sec	1.689
Knot	kilometre/hour	1.8532
Knot	metre/minute	30.883 261 27
Knot (*International*)	metre/sec	0.5144
Knot	mile (*nautical*) /hour	1.0
Knot	mile (*statute*)/hour	1.151
Knot	yard/hour	2.027×10^3

** The British Admirality uses the round figure of 6080 feet.*

L

To Convert	*into*	*multiply by*
Lambert (*luminance, photometric brightness,* π^1 *cd/cm*^{2})	apostilb	1.0×10^4
Lambert	cd/cm^2	0.318 268 62
Lambert	cd/in.^{2}	2.053 388 09
Lambert	cd/metre^2	$3.183\ 268\ 62 \times 10^3$
Lambert	nit	$(1/\pi) \times 10^4$
Lambert	stilb	0.318
Langley (*solar radiant exposure, gram-calorie/cm*^{2})	joule/metre^2	4.184×10^4
Langley/min (*solar irradiance,insolation*)	watt/metre^2	$6.97\overline{3}^* \times 10^2$
League	mile (*approx.*)	3.0
Leap year	day	3.66×10^2
Light year	kilometre	$9.460\ 91 \times 10^{12}$
Light year	mile	5.9×10^{12}
Line	weber (*Wb*)	1.0×10^{-8}
Line/cm^2	gauss	1.0
Line/in.^{2}	gauss	0.155
Line/in.^{2}	weber/cm^2	1.55×10^{-9}
Line/in.^{2}	weber/in.^{2}	1.0×10^{-8}
Line/in.^{2}	weber/metre^2	1.55×10^{-5}
Link	chain	1.0×10^{-2}
Link (*surveyor's*)	ft	0.66
Link	furlong	1.0×10^{-3}
Link (*engineer's*)	in.	12.0
Link (*surveyor's*)	in.	7.92
Link	rod	4.0×10^{-2}

* *symbol — over a digit indicate recurring decimal digit*

To Convert	*into*	*multiply by*
Link^2 (*Gunter's*)	acre	1.0×10^{-5}
Litre	bushel (*US; dry*)	2.838×10^{-2}
Litre	centilitre	1.0×10^{2}
Litre	cm^3	$1.000\ 028 \times 10^{3}$
Litre	decilitre	10.0
Litre	dekalitre	0.1
Litre	ft^3	$3.531\ 56 \times 10^{-2}$
Litre	gallon (UK)	0.219 976
Litre	gallon (*US; liquid*)	0.264 178
Litre	gill	7.039 031 43
Litre	hectolitre	1.0×10^{-2}
Litre	in.^{3}	61. 025 454
Litre	metre^3	$1.000\ 028 \times 10^{-3}$
Litre	ounce (*US; liquid*)	33.814 023 34
Litre	peck (*UK*)	0.109 988
Litre	peck (*US*)	0.113 512 76
Litre	pint (*UK; liquid*)	1.759 804 31
Litre	pint (*US; liquid*)	2.113 42
Litre	pound	2.2046
Litre	quart (*US; liquid*)	1.057
Litre	quart (*dry*)	0.9081
Litre	yard^3	$1.307\ 98 \times 10^{-3}$
Litre-atmosphere	Btu	$9.607\ 071 \times 10^{-2}$
Litre/hectare	gallon (*UK*)/acre	$8.902\ 21 \times 10^{-2}$
Litre/hectare	gallon (*US*)/acre	0.106 906 62
Litre/hour	gallon (*UK*)/hour	0.219 973 6
Litre/hour	gallon (*US*)/hour	0.264 172 04
Litre/kg	ft^3/lb	$1.601\ 89 \times 10^{-2}$
Litre/kg	gallon(*UK*)/lb	9.9779×10^{-2}
Litre/kg	in.^{3}/lb	27. 6807

To Convert	*into*	*multiply by*
Litre/kg	metre³/kg	$1.000\ 028 \times 10^{-3}$
Litre/km	gallon(*UK*)/mile	0.354 016
Litre/min	ft³/sec	$5.885\ 93 \times 10^{-4}$
Litre/min	gallon/min	0.264 172 04
Litre/min	gallon/sec	4.403×10^{-3}
Litre/sec	ft³/min	2.119 093 03
Litre/sec	ft³/sec	$3.531\ 56 \times 10^{-2}$
Litre/sec	gallon (*UK*)/min	13.198 416 19
Litre/sec	gallon (*US*)/min	15.85
Litre/sec	metre³/hour	3.600 109
Litre/sec	million gallon (*US*)/day	2.28×10^{-2}
Litre/sec	yard³/min	$7.849\ 294 \times 10^{-2}$
Lumen	spherical candle power	7.958×10^{-2}
Lumen	watt	1.496×10^{-3}
Lumen/ft²	foot-candle	1.0
Lumen/ft²	lumen/metre² (*lux, lx*)	10.7639
Lumen/metre² (*lx*)	lumen/ft²	9.29×10^{-2}
Lux	foot-candle	9.29×10^{-2}
Lux	phot (*illuminance, lm/cm²*)	1.0×10^{-4}

M

To Convert	*into*	*multiply by*
Manual labor (one hour)*	kWh	0.06
Maun**	hundredweight (*cwt*)	0.734 69
Maund	kilogram	37.3242
Maund	pound	82.286
Maund	pound (*apothecaries*)	1.0×10^{2}
Maund	seer**	40.0
Maund	ton (*long*)	$3.673\,47 \times 10^{-2}$
Maund	ton (*short*)	$4.114\,29 \times 10^{-2}$
Maxwell (*magnetic flux in coil rating*)	cgse	$3.\overline{3}$*** $\times 10^{-11}$
Maxwell	kiloline	1.0×10^{-3}
Maxwell	megaline	1.0×10^{-6}
Maxwell	weber	1.0×10^{-8}
Megadyne (*MDyn*)	kgf	1.019 72
Megadyne	pound-f	2.248 09
Megadyne	poundal	72.33
Megajoule	hp-hour	0.372 506
Megajoule	kWh	0.27***
Megajoule	therm	$9.478\,67 \times 10^{-3}$
Megaline	Maxwell	1.0×10^{6}
Megaohm	microhm	1.0×10^{12}
Megaohm	ohm	1.0×10^{6}
Metre	angstrom unit (Å)	1.0×10^{10}
Metre	bolt (*US; cloth*)	$2.734\,033 \times 10^{-2}$
Metre	centimetre	1.0×10^{2}
Metre	chain	$4.970\,97 \times 10^{-2}$

* *One hour manual labour ≃ 0.06 k Wh*

** *an Indian unit of weight which varies greatly in different places*

*** *symbol — over a digit indicate recurring decimal digit*

To Convert	*into*	*multiply by*
Metre	decimetre	10.0
Metre	dekametre	0.1
Metre	fathom	0.546 81
Metre	fermi	1.0×10^{15}
Metre	ft	3.280 839 895
Metre	hectometre	1.0×10^{-2}
Metre	in.	39.3701
Metre	kilometre	1.0×10^{-3}
Metre	mil	3.937×10^{4}
Metre	mile (*nautical*)	5.396×10^{-4}
Metre	mile (*statute*)	6.214×10^{-4}
Metre	millimetre	1.0×10^{3}
Metre	millimicron	1.0×10^{9}
Metre	rod	0.198 839
Metre	varas	1.179
Metre	yard	1.093 613 298
Metre/minute	cm/sec	$1.6\bar{6}$*
Metre/minute	ft/minute	3.281
Metre/minute	ft/sec	5.468×10^{-2}
Metre/minute	km/hour	6.0×10^{-2}
Metre/minute	knot**	3.238×10^{-2}
Metre/minute	mile/hour	3.728×10^{-2}
Metre/sec	ft/minute	1.968×10^{2}
Metre/sec	ft/sec	3.280 84
Metre/sec	km/hour	3.6
Metre/sec	km/minute	6.0×10^{-2}
Metre/sec	knot (*International*)	1.944 012 44
Metre/sec	mile/hour	2.236 94
Metre/sec	mile (*nautical, International*)/hour	1.943 84

* *symbol — over a digit indicate recurring decimal digit*

** *One nautical mile (International)/hour=1 knot*

To Convert	*into*	*multiply by*
Metre/sec	mile (*nautical,* UK)/hour	1.9426
Metre/sec	mile/minute	3.728×10^{-2}
Metre/sec^2	free fall (*standard*)	0.101 967 98
Metre/sec/sec	cm/sec/sec	1.0×10^{2}
Metre/sec/sec	ft/sec/sec	3.281
Metre/sec/sec	km/hour/sec	3.6
Metre/sec/sec	mile/hour/sec	2.237
Metre-kilogram	cm-dyne	$9.806\,65 \times 10^{7}$
Metre-kilogram	cm-gram	1.0×10^{5}
Metre-kilogram	pound-ft	7.233
Metre-kilogramf	erg	$9.806\,65 \times 10^{-7}$
Metre-kilogramf	joule	9.806 65
Metre-kilogramf/sec	atmosphere	9.67×10^{-2}
Metre-kilogramf/sec	cheval vapeur	$1.3\bar{3}$* $\times 10^{-2}$
Metre-kilogramf/sec	erg/sec	$9.806\,65 \times 10^{-7}$
Metre-kilogramf/sec	ft-poundf/sec	7.233
Metre-kilogramf/sec	horsepower	$1.315\,09 \times 10^{-2}$
Metre-kilogramf/sec	kilowatt	$9.806\,65 \times 10^{-3}$
Metre-kilogramf/sec	watt	9.806 65
Metre-newton-sec/metre^2	pound/ft-hour (*viscosity*)	2.415 458 94
Metre of water	atmosphere	$9.671\,18 \times 10^{-2}$
Metre of water	in. of mercury	2.89
Metre of water	kilogram	9.806 65
Metre of water	kilogram/cm^2	0.1
Metre of water	pound/in.^{2}	1.421
Metre^2	acre	2.471×10^{-4}
Metre^2	are	1.0×10^{-2}
Metre^2	barn	1.0 × 1028

* *symbol — over a digit indicate recurring decimal digit*

To Convert	*into*	*multiply by*
Metre^2	bigha	$3.953\ 686\ 103 \times 10^{-4}$
Metre^2	centare (*centiare*)	1.0
Metre^2	circular mil	1.973×10^{9}
Metre^2	cm^2	1.0×10^{4}
Metre^2	ft^2	10. 763 910 42
Metre^2	hectare	1.0×10^{-4}
Metre^2	in.^{2}	1.55×10^{3}
Metre^2	km^2	1.0×10^{-6}
Metre^2	mile^2	3.861×10^{-7}
Metre^2	mm^2	1.0×10^{6}
Metre^2	yard^2	1.195 990 046
Metre^2/hour	ft^2/hour (*thermal diffusivity*)	10.763 910 42
Metre^2/hour	stoke	$2.7\overline{7}$*
Metre^2/litre	yard^2/gallon (*UK; suface coating*)	5.437 737 9
Metre^2/sec	centistoke	1.0×10^{6}
Metre^2/sec	stoke	1.0×10^{4}
Metre^3	acre-ft	$8.107\ 130\ 9 \times 10^{-4}$
Metre^3	barrel (*liquid; US*)	6.2898
Metre^3	bushel (*UK*)	27.4962
Metre^3	bushel (*dry; US*)	28.3774
Metre^3	cm^3	1.0×10^{6}
Metre^3	ft^3	35.314 666 72
Metre^3	gallon (*UK*)	2.2×10^{2}
Metre3	gallon (*US; dry*)	$2.270\ 147\ 56 \times 10^{2}$
Metre^3	gallon (*US; liquid*)	2.6418×10^{2}
Metre^3	in.^{3}	$6.102\ 338 \times 10^{4}$
Metre^3	kilolitre	0.999 97

* *symbol — over a digit indicate recurring decimal digit*

To Convert	*into*	*multiply by*
Metre^3	litre	9.9997×10^2
Metre^3	peck (US)	$1.135\,073\,78 \times 10^2$
Metre^3	pint (*dry; US*)	1.8162×10^3
Metre^3	pint (*liquid; US*)	2.113×10^3
Metre^3 (*water*)	pound	2.204×10^3
Metre^3	quart (*liquid; US*)	1.057×10^3
Metre^3	table spoon	$6.761\,325 \times 10^4$
Metre^3	tea spoon	$2.028\,809 \times 10^5$
Metre^3	yard^3	1.307 950 619
Metre^3/hectare (*fodder yield*)	ft^3/acre	14.2916
Metre^3/hour	ft^3/minute	0.588 581 52
Metre^3/hour	ft^3/sec	9.81×10^{-3}
Metre^3/hour	gallon(US)/minute	4.403
Metre^3/hour	litre/sec	0.277 769 4
Metre^3/hour	million gallon (US)/day	6.34×10^{-3}
Metre^3/kg	ft^3/pound	16.0185
Metre^3/kg	gallon(UK)/pound	99.776
Metre^3/kg	in.^{3}/pound	$2.767\,99 \times 10^4$
Metre^3/kg	litre/kg	9.9997×10^2
Metre^3/minute	ft^3/sec	0.588 577 71
Metre^3/Pa-sec	cm^3/dyne-sec	1.0×10^5
Metre^3/sec	gallon (*US; liquid*) /min	$1.585\,037\,2 \times 10^4$
Metric minute	degree	$9.009\,01 \times 10^{-3}$
Metric minute	radian	$1.569\,858\,7 \times 10^{-4}$
Metric second	degree	$9.009\,01 \times 10^{-5}$
Metric second	radian	$1.569\,858\,7 \times 10^{-6}$
Metric technical weight (*mass*)	kilogram	9.806 65

To Convert	*into*	*multiply by*
Metric technical weight (*mass*)	pound (*avoirdupois*)	21.62
Metric technical weight (*avoirdupois*)	slug	0.671 97
Mho (*conductance, susceptance, admittance*)	siemen (*S*)	1.0
Mho/cm (*conductivity*)	siemen/metre	1.0×10^2
Mho/ft	siemen/metre	3.280 84
Mho (m, g) (*electric conductivity, mass density*)	abmho/cm	1.0×10^{-5} δ*
Mho (m, g)	mho (mil-ft)	1.662×10^{-3} δ*
Mho (mil-ft) (*electric conductivity, volume basis*)	abmho/cm	6.015×10^{-3}
Mho (mil-ft)	mho (m, g)	$6.015/\delta^* \times 10^2$
Micro-in. (*surface texture roughness*)	micrometre	2.54×10^{-2}
Micro-in. (*radiation wavelength, optical measurement, interference*)	nanometre (*nm*)	25.4
Micrometre	micro-in.	39.370 078 74
Micron (μ)	angstrom unit (*Å*)	1.0×10^4
Micron	mil	$3.937\,008 \times 10^{-2}$
Micron	nanometre	1.0×10^5
Micro ohm	megaohm	1.0×10^{-12}
Micro ohm	ohm	1.0×10^{-6}
Micro ohm-cm (*electrical resistivity, volume basis*)	abohm-cm	1.0×10^3
Micro ohm-cm	microhm-in.	0.3937

* δ = density, g/cm³

To Convert	*into*	*multiply by*
Micro ohm-cm	ohm (mil-ft)	6.015
Micro ohm-cm (*electrical resistivity, mass density basis*)	ohm (m,g)	$1.0 \times 10^{2}\ \delta$*
Micro ohm-in. (*electrical resistivity, volume basis*)	abohm-cm	2.54×10^{3}
Micro ohm-in.	micro ohm-cm	2.54
Micro ohm-in.	ohm (mil,ft)	15.28
Micro ohm-in. (*electrical resistivity, mass density basis*)	ohm (m,g)	$2.54 \times 10^{-2}\ \delta$*
Mil	centimetre	2.54×10^{-3}
Mil	ft	$8.3\overline{3}$** $\times 10^{-5}$
Mil	in.	1.0×10^{-3}
Mil	kilometre	2.54×10^{-8}
Mil	metre	2.54×10^{-5}
Mil	micron	25.4
Mil	mm	2.54×10^{-2}
Mil	yard	2.778×10^{-5}
Mil (*angular*)	degree	5.625×10^{-2}
Mil (*angular*)	minute (′)	3.375
Mil (*angular*)	radian	9.8175×10^{-4}
Mil (*angular*)	second (″)	2.025×10^{2}
Mil^2	cm^2	6.452×10^{-6}
Mil^2	circular mil	1.273
Mil^2	$in.^2$	1.0×10^{-6}
Mil-ft	$in.^3$	9.425×10^{-6}
Mile	cable length	$7.33\overline{3}$**
Mile (*nautical, UK*)	cable length	$8.4\overline{4}$**

* δ = density, g/cm^3

** *symbol — over a digit indicate recurring decimal digit*

To Convert	*into*	*multiply by*
Mile (*nautical, International*)	cable length	8.439
Mile (*nautical*)	cm	1.853×10^5
Mile (*statute*)	cm	$1.609\ 344 \times 10^5$
Mile (*nautical*)	ft	$6.076\ 115\ 7 \times 10^3$
Mile (*statute*)	ft	5.28×10^3
Mile	furlong	8.0
Mile (*nautical, International*)	furlong	9.2062
Mile (*nautical, UK*)	furlong	9.2121
Mile (*nautical*)	in.	$7.291\ 338\ 8 \times 10^4$
Mile (*statute*)	in.	6.336×10^4
Mile (*nautical, International*)	kilometre	1.853 184
Mile (*nautical; UK*)	kilometre	1.853 190 27
Mile (*statute*)	kilometre	1.609 344
Mile (*approx.*)	league	$0.3\overline{3}$*
Mile	light year	$1.694\ 915\ 3\times10^{-13}$
Mile (*nautical*)	metre	$1.853\ 184 \times 10^3$
Mile (*nautical, US*)	metre	1.852×10^3
Mile (*statute*)	metre	$1.609\ 269\ 4 \times 10^3$
Mile (*nautical, International*)	mile (*nautical; International*)	0.999 36
Mile (*nautical, UK*)	mile (*nautical; International*)	1.000 64
Mile (*nautical*)	mile (*statute*)	1.150 779 5
Mile (*statute*)	mile (*nautical*)	0.868 976 24
Mile	millimetre	$1.609\ 269\ 4 \times 10^6$
Mile	parsec	$5.263\ 157\ 9\times10^{-14}$
Mile (*nautical*)	yard	$2.026\ 67 \times 10^3$

***** *symbol* — *over a digit indicate recurring decimal digit***

To Convert	*into*	*multiply by*
Mile (*nautical, International*)	yard	$2.025\ 37 \times 10^{3}$
Mile (*statute*)	yard	1.76×10^{3}
Mile/gallon (*fuel efficiency, US*)	kilometre/litre	0.415 143 7
Mile/gallon (*UK*)	kilometre/litre	0.354 016
Mile/hour	cm/sec	44.7
Mile/hour	ft/minute	88.0
Mile/hour	ft/sec	1.467
Mile/hour	km/hour	1.609 344
Mile/hour	km/minute	2.682×10^{-2}
Mile (*nautical*)/hour	knot	1.0
Mile/hour	knot	0.868 976 24
Mile/hour	metre/minute	26.82
Mile (*statute, US*)/hour	metre/sec	0.447
Mile (*nautical, International*)/hour	metre/sec	0.514 445 63
Mile (*nautical, UK*)/hour	metre/sec	0.514 774
Mile/hour	mile/minute	$0.1\overline{6}6$*
Mile/hour/sec	cm/sec/sec	44.7
Mile/hour/sec	ft/sec/sec	1.467
Mile/hour/sec	km/hour/sec	1.609 344
Mile/hour/sec	metre/sec/sec	0.447
Mile/minute	cm/sec	2.682×10^{3}
Mile/minute	ft/sec	88.0
Mile/minute	km/minute	1.609 344
Mile/minute	metre/sec	26.824 034 33
Mile/minute	mile/hour	60.0
Mile^2	acre	6.4×10^{2}
Mile^2	cm^2	2.59×10^{10}

** symbol — over a digit indicate recurring decimal digit*

To Convert	*into*	*multiply by*
Mile^2	ft^2	2.788×10^7
Mile^2	hectare	$2.589\,99 \times 10^2$
Mile^2	in.^{2}	4.015×10^9
Mile^2	kilometre^2	2.589 998
Mile^2	metre^2	$2.589\,998 \times 10^6$
Mile^2	millimetre^2	$2.590\,002\,6 \times 10^{12}$
Mile^2	rod^2	1.024×10^5
Mile^2	rood	2.56×10^3
Mile^2	yard^2	3.0976×10^6
Millibar	atmosphere	9.8692×10^{-4}
Millibar	in. of mercury	2.953×10^{-2}
Millibar	in. of water	0.401 463
Millibar	kgf/metre^2	10.1972
Millibar	mm of mercury	0.750 06
Millibar	pascal	1.0×10^2
Millibar	pound/ft^2	2.088 54
Milligram	grain	$1.543\,236 \times 10^{-2}$
Milligram	gram	1.0×10^{-3}
Milligram/litre	grain/ft^3	0.436 983
Milligram/litre	grain/gallon(*UK*)	7.0155×10^{-2}
Milligram/litre	grain/gallon(*US*)	$5.841\,61 \times 10^{-2}$
Milligram/litre	gram/metre^3	0.999 97
Milligram/litre	ounce (*avoirdu-pois*)/ft^3	$9.988\,213\,9 \times 10^{-4}$
Milligram/litre	ounce (*avoirdup-ois*)/gallon (*UK*)	$1.603\,540\,6 \times 10^{-4}$
Milligram/litre	parts per million (*ppm*)	1.0
Milli-henrie	henrie	1.0×10^{-3}
Millilitre	cm^3	1.000 03
Millilitre	gill (*UK*)	7.0392×10^{-3}

To Convert	*into*	*multiply by*
Millilitre	$in.^3$	6.1025×10^{-2}
Millilitre (*water*)	kilogram	9.9997×10^{2}
Millilitre	minim (*UK*)	16.8941
Millilitre	minim (*US*)	16.2311
Millilitre	ounce (*UK*)	$3.519\ 61 \times 10^{-2}$
Millilitre	ounce (*US*)	$3.381\ 48 \times 10^{-2}$
Millilitre/sec	ounce/sec	$3.381\ 402 \times 10^{-2}$
Millimetre	cm	0.1
Millimetre	ft	3.281×10^{-3}
Millimetre	in.	3.937×10^{-2}
Millimetre	kilometre	1.0×10^{-6}
Millimetre	metre	1.0×10^{-3}
Millimetre	mil	39.37
Millimetre	mile	6.214×10^{-7}
Millimetre	yard	1.094×10^{-3}
Millimetre of mercury	atmosphere	$1.315\ 79 \times 10^{-3}$
Millimetre of mercury	$dyne/cm^2$	$1.3\overline{3}^* \times 10^{3}$
Millimetre of mercury	in. of water	0.535 24
Millimetre of mercury	$kgf/metre^2$	13.5951
Millimetre of mercury	millibar	1.333 22
Millimetre of mercury (*at 0ºC*)	pascal (*Pa*)	$1.3\overline{3}^* \times 10^{2}$
Millimetre of mercury	$poundf/ft^2$	2.7845
Millimetre of mercury	$poundf/in.^2$	1.936×10^{-2}
$Millimetre^2$	cm^2	1.0×10^{-2}
$Millimetre^2$	circular mil	1.973×10^{3}
$Millimetre^2$	ft^2	1.076×10^{-5}
$Millimetre^2$	$in.^2$	1.55×10^{-3}
$Millimetre^2$	$kilometre^2$	1.0×10^{-12}
$Millimetre^2$	$metre^2$	1.0×10^{-6}

** symbol — over a digit indicate recurring decimal digit*

To Convert	*into*	*multiply by*
Millimetre^2	mile^2	3.861×10^{-13}
Millimetre^2	yard^2	1.196×10^{-6}
Millimetre^4	in.^{4}	$2.402\ 509\ 8 \times 10^{-6}$
Millimicron	metre	1.0×10^{-9}
Millinewton/metre	dyne/cm	1.0
Millinewton-metre	ounce force-in.	0.141 611 93
Million gallon/day	ft^3/sec	1.547 23
Miner's inch*	ft^3/minute	1.53
Minim (*UK*)	cm^3	5.9194×10^{-2}
Minim (*US*)	cm^3	6.1612×10^{-2}
Minim (*UK*)	gill	$4.16\overline{6}$** $\times 10^{-4}$
Minim (*US*)	gill	$4.336\ 870\ 8 \times 10^{-4}$
Minim (*UK*)	in.^{3}	$3.612\ 23 \times 10^{-3}$
Minim (*US*)	in.^{3}	$3.759\ 79 \times 10^{-3}$
Minim (*UK*)	millilitre	5.9192×10^{-2}
Minim (*US*)	millilitre	6.161×10^{-2}
Minim (*US*)	minim (*UK*)	1.040 85
Minim (*US*)	ounce (*US*)	$2.083\ 33 \times 10^{-3}$
Minute (*angular*)	degree	$1.\overline{6}$** $\times 10^{-2}$
Minute (*angular*)	mil (*angular*)	0.296 296
Minute (*angular*)	quadrant	$1.851\ 85 \times 10^{-4}$
Minute (*angular*)	radian	$2.908\ 88 \times 10^{-4}$
Minute (*angular*)	revolution	4.63×10^{-5}
Minute (*angular*)	second (", *angular*)	60.0
Minute (*time*)	arc	37.5
Minute (*time*)	hour	$1.6\overline{6}$** $\times 10^{-2}$
Minute (*time*)	second	60.0
Moisture content (*dry basis, D*),%	moisture content (*wet basis, W*),%	W,%=100D/(100+D)

* *fluid flow measure (definition under section 2.9.1)*

** *symbol — over a digit indicate recurring decimal digit*

To Convert	*into*	*multiply by*
Moisture content (*wet basis, W*),%	moisture content (*dry basis, D*),%	D,%=100W/(100-W)
Musical interval (*major tone sixth, A*)	cent (*natural scale*)	8.84×10^2
Musical interval (*major tone sixth, A*)	cent (*tempered*)	9.0×10^2
Musical interval (*major tone sixth, A*)	savart (*natural scale*)	2.219×10^2
Musical interval (*major tone sixth, A*)	savart (*tempered scale*)	2.258×10^2
Musical interval (*major tone sixth, A*)	unison (*C tone, natural scale*)	5/3
Musical interval (*major tone, D*)	cent (*natural scale*)	2.04×10^2
Musical interval (*major tone, D*)	cent (*tempered*)	2.0×10^2
Musical interval (*major tone, D*)	savart (*natural scale*)	51.1
Musical interval (*major tone, D*)	savart (*tempered*)	50.2
Musical interval (*major tone, D*)	unison (*C tone, natural scale*)	9/8
Musical interval (*major tone third, E*)	cent (*natural scale*)	3.86×10^2
Musical interval (*major tone third, E*	cent (*tempered*)	4.0×10^2
Musical interval (*major tone third, E*	savart (*natural* scale)	96.9
Musical interval (*major tone third, E*	savart (*tempered*)	1.004×10^2
Musical interval (*major tone third, E*	unison (*C tone, natural scale*)	5/4

To Convert	*into*	*multiply by*
Musical interval (*fourth, F tone*)	cent (*natural scale*)	4.98×10^2
Musical interval (*fourth, F tone*)	cent (*tempered*)	5.0×10^2
Musical interval (*fourth, F tone*)	savart (*natural scale*)	1.25×10^2
Musical interval (*fourth, F tone*)	savart (*tempered*)	1.254×10^2
Musical interval (*fourth, F tone*)	unison (*C tone, natural scale*)	4/3
Musical interval (*fifth, G tone*)	cent (*natural scale*)	7.02×10^2
Musical interval (*fifth, G tone*)	cent (*tempered*)	7.0×10^2
Musical interval (*fifth, G tone*)	savart (*natural scale*)	1.761×10^2
Musical interval (*fifth, G tone*)	savart (*tempered*)	1.756×10^2
Musical interval (*fifth, G tone*)	unison (*C tone, natural scale*)	3/2
Myria gram	kilogram	10.0
Myria metre	kilometre	10.0
Myria watt	kilowatt	10.0

N

To Convert	*into*	*multiply by*
Nail	in.	2.25
Nail	quarter (*length, cloth*)	0.25
Nanometre (*nm*)	angstrom	0.1
Nanometre	micro-in.	$3.937\ 008 \times 10^{-2}$
Nanometre	micron (μ)	1.0×10^{-5}
Neper	decible	8.686
Newton	dyne	1.0×10^{5}
Newton (*N*)	kilogram force	0.101 971 62
Newton	ounce force	3.596 942 46
Newton	pound force	0.224 808 92
Newton/metre	pound/ft	$6.852\ 178 \times 10^{-2}$
Newton-metre (N-m)	kilogramf-cm	10.197 162 13
Newton-metre	poundf-ft	0.737 562 12
Newton-metre	poundf-in.	8.850 748 07
Newton/metre^2	pascal	1.0
Newton/mm	poundf/in.	5.710 148 3
Newton-sec/metre	pound-sec/ft	$6.852\ 178 \times 10^{-2}$
Nit (*luminance*)	apostilb	3.14
Nit	lambert	3.14×10^{-4}
Nit	stilb	1.0×10^{-4}

O

To Convert	*into*	*multiply by*
Octave (*C tone, musical interval*)	cent (*natural*)	1.2×10^3
Octave	cent (*tempered*)	1.2×10^3
Octave	savart (*natural*)	3.01×10^2
Octave	savart (*tempered*)	3.01×10^2
Octave	unison (*C tone, natural*)	2.0
Oersted (*strength, magnetic field, Oe*)	ampere/metre	79.577 47
Oersted	cgse	3.0×10^{10}
Ohm (*electric resistance*)	abohm	1.0×10^9
Ohm	cgse	1.11×10^{-12}
Ohm	cgsm	1.0×10^9
Ohm	megaohm	1.0×10^{-6}
Ohm	microhm	1.0×10^6
Ohm (*International*)	ohm (*absolute*)	1.0005
Ohm	statohm	1.112×10^{-12}
Ohm(m,g) (*electrical resistivity, mass basis*)	abohm-cm	$1.0 \times 105/\delta^*$
Ohm (m,g)	microhm-cm	$1.0 \times 102/\delta^*$
Ohm (m,g)	microhm-in.	$39.37/\delta^*$
Ohm (m,g)	ohm(mil,foot) (*volume basis*)	$6.015/\delta^* \times 10^2$
Ohm (mil,ft) (*electrical resistivity, volume basis*)	abohm-cm	1.662×10^2
Ohm (mil,ft)	microhm-cm	0.1662
Ohm (mil,ft)	microhm-in.	6.545×10^{-2}
Ohm (mil,ft)	ohm (m,g)	$1.662 \times 10^{-3}\ \delta^*$

* δ = mass density, g/cm^3

To Convert	*into*	*multiply by*
Ohm-metre (*resistivity, electric*)	cgse	1.11×10^{-10}
Ohm-metre	cgsm	1.0×10^{11}
Ohm-metre	ohm-mm²/metre	1.0×10^{6}
Ohm-millimetre²/metre	cgse	1.11×10^{-16}
Ohm-millimetre²/metre	cgsm	1.0×10^{5}
Ohm-millimetre²/metre	ohm-metre	1.0×10^{-6}
Ounce (*apothecarie's or troy, oz*)	carat (*metric*)	$1.555\,17 \times 10^{2}$
Ounce (*apothecarie's or troy, oz*)	drachm (*apothecarie's*)	8.0
Ounce (*apothecarie's or troy, oz*)	dram (*avoirdupois*)	17.5543
Ounce (*apothecarie's or troy, oz*)	grain	4.8×10^{2}
Ounce (*apothecarie's or troy, oz*)	gram	31.103 481
Ounce (*apothecarie's or troy, oz*)	ounce (*avoir-dupois*)	1.097 14
Ounce (*apothecarie's or troy, oz*)	pennyweight (*troy*)	20.0
Ounce (*apothecarie's or troy, oz*)	pound (*troy*)	8.333×10^{-2}
Ounce (*apothecarie's or troy, oz*)	tola	$2.6\overline{6}$*
Ounce (*avoirdupois, oz*)	carat (*metric*)	$1.417\,474\,6 \times 10^{2}$
Ounce (*avoirdupois, oz*)	dram (*avoirdupois*)	16.0
Ounce (*avoirdupois, oz*)	drachm	7.291 664 39
Ounce (*avoirdupois, oz*)	grain	4.375×10^{2}
Ounce (*avoirdupois, oz*)	gram	28.349 523 13
Ounce (*avoirdupois, oz*)	hundred weight (*short*)	6.25×10^{-4}

** symbol — over a digit indicate recurring decimal digit*

To Convert	*into*	*multiply by*
Ounce (*fluid, oz*)	gallon (*US*)	1/128
Ounce (*US, fluid, oz*)	centilitre	2.957 352 9
Ounce (*US, oz*)	cm^3	29.5737
Ounce (*UK*)	cm^3	28.413 046 13
Ounce (*UK, fluid*)	gill	0.2
Ounce (*US, liquid*)	gill	0.208 169 83
Ounce (*US; fluid, oz*)	$in.^3$	1.8047
Ounce (*UK, fluid*)	$in.^3$	1.733 883 55
Ounce (*US; fluid, oz*)	litre	$2.957\ 352\ 9 \times 10^{-2}$
Ounce (*US; fluid, oz*)	millilitre	29.573 529 56
Ounce (*UK*)	millilitre	28.412 238 86
Ounce (*US; fluid, oz*)	minim (*US*)	4.8×10^{2}
Ounce (*US; fluid, oz*)	ounce (*UK, fluid*)	1.040 85
Ounce (*UK, fluid*)	ounce (*US, fluid*)	0.960 753 23
Ounce (*avoirdupois*)	ounce (*troy, apothecarie's*)	0.911 461
Ounce (*fluid*)	pint (*US*)	6.25×10^{-2}
Ounce	pound	6.25×10^{-2}
Ounce (*troy*)	pound (*avoirdupois*)	$6.857\ 153 \times 10^{-2}$
Ounce (*avoirdupois*)	pound (*troy*)	$7.595\ 494 \times 10^{-2}$
Ounce	quart (*fluid*)	3.125×10^{-2}
Ounce (*avoirdupois*)	tola	2.430 553 02
Ounce	tonne (*long*)	2.79×10^{-5}
Ounce	tonne (*metric*)	2.835×10^{-5}
Ounce	tonne (*short*)	3.125×10^{-5}
Ounce (*troy*)	tonne (*short*)	$3.42\ 857\ 15 \times 10^{-5}$
Ounce force (*ozf*)	newton (*N*)	0.278 013 9
Ounce/ft^2 (*fabric, surface coating*)	gram/$metre^2$	$3.051\ 517 \times 10^{2}$
Ounce (avoirdupois)/ft^3	grain/ft^3	4.375×10^{2}

Ounce (avoirdupois)/ft^3	grain/gallon (UK)	70.238
Ounce (avoirdupois)/ft^3	gram/metre^3	$1.001\ 15 \times 10^3$
Ounce (avoirdupois)/ft^3	milligram/litre	$1.001\ 18 \times 10^3$
Ounce (avoirdupois)/ft^3	ounce (avoirdupois)/gallon (UK)	0.160 544
Ounce (avoirdupois)/gallon (UK)	grain/ft^3	$2.725\ 12 \times 10^3$
Ounce (avoirdupois)/gallon (UK)	grain/gallon (UK)	4.375×10^2
Ounce (avoirdupois)/gallon (UK)	gram/metre^3	6.236×10^3
Ounce (avoirdupois)/gallon (UK)	milligram/litre	6.2362×10^3
Ounce (avoirdupois)/gallon (UK)	ounce (*avoirdupois*)/ft^3	6.2288
Ounce force-in. (*light torque*)	millinewton-metre	7.061 552
Ounce-in. (*moment of mass*)	gram-metre	0.720 077 8
Ounce/in.^{2}	dyne/cm^2	4.309×10^3
Ounce/in.^{2}	inch (*of water, 4°C*)	1.729 804 53
Ounce/in.^{2}	pound/in.^{2}	6.25×10^{-2}
Ounce/sec	millilitre/sec	29.573 53
Ounce/yard^2 (*fabric, surface coating*)	gram/metre^2	33.905 75

P

To Convert	*into*	*multiply by*
Palm	in.	3.0
Parsec	kilometre	3.084×10^{13}
Parsec	mile	1.9×10^{13}
Parts per million (*ppm*)	grain/gallon(*UK*)	7.016×10^{-2}
Parts per million	grain/gallon(*US*)	5.84×10^{-2}
Parts per million	gram/litre	1.0×10^{-3}
Parts per million	milligram/litre	1.0
Parts per million	pound/million gallon	8.345
Pascal (*Pa*)	atmosphere	8.7×10^{-7}
Pascal	bar (*hecto pieze*)	1.0×10^{-5}
Pascal	cm (*of mercury at 0⁰C*)	7.5×10^{-4}
Pascal	cm (*of water at 4⁰C*)	$1.019\,784 \times 10^{-2}$
Pascal	dyne/cm^2	10.0
Pascal	ft (*of water*)	$3.345\,600\,5 \times 10^{-4}$
Pascal	Inch (*of mercury; 32⁰F*)	$2.952\,998 \times 10^{-4}$
Pascal	Inch (*of mercury; 60⁰F*)	$2.961\,339\,7 \times 10^{-4}$
Pascal	Inch (*of water; 4⁰C*)	40.147 421 33
Pascal	Inch (*of water; 60⁰F*)	40.186 465 2
Pascal	kgf/metre^2	0.102
Pascal	millibar	1.0×10^{-2}
Pascal	mm of mercury	7.5×10^{-3}
Pascal	newton/metre^2	1.0
Pascal	pieze	1.0×10^{-3}

To Convert	*into*	*multiply by*
Pascal	poundf/in.^{2}	$1.450\ 377\ 4 \times 10^{-4}$
Pascal-sec	centipoise (*viscosity, dynamic*)	1.0×10^3
Pascal-sec	dyne-sec/cm^5	1.0×10^{-5}
Pascal-sec	poise	10.0
Pascal-sec	pound mass/ft-sec	0.672 043 01
Pascal-sec	poundf-sec/ft^2	$2.088\ 542 \times 10^{-2}$
Pascal-sec	slug/ft-sec	$2.088\ 555 \times 10^{-2}$
Pascal-sec/metre	dyne-sec/cm^5 (*accoustic impedance/resistance /reactance*)	1.0×10^{-5}
Peck (*UK*)	in.^{3}	5.546×10^2
Peck (*UK*)	litre	9.091 901
Peck (*US*)	bushel	0.25
Peck (*US*)	in.^{3}	$5.376\ 05 \times 10^2$
Peck (*US*)	litre	8.809 582
Peck (*US*)	metre^3	8.81×10^{-3}
Peck (*US*)	quart (*dry*)	8.0
Pennyweight (*troy*)	grain	24.0
Pennyweight (*troy*)	gram	1.555 17
Pennyweight (*troy*)	ounce (*troy*)	5.0×10^{-2}
Pennyweight (*troy*)	pound (*troy*)	4.1667×10^{-3}
Percent grade	ft/100 ft	1.0
Perch	rod	1.0
Perch of stone* (*volume*)	ft^3	24.75
Perm (*0°C*)	kg/pascal-sec-metre^2	5.721×10^{-11}
Phot (*illum nance, lm/cm^2*)	lux (*lx*)	1.0×10^4
Pieze	atmosphere	$9.900\ 99 \times 10^{-3}$

* *a perch of stone equals to a quantity ½ foot thick, 1 foot high ad 16½ feet long = 24 ¾ feet^3*

To Convert	*into*	*multiply by*
Pieze	bar	1.0×10^{-2}
Pieze	dyne	1.0×10^{4}
Pieze	kilogramf/cm²	$1.019\,716 \times 10^{-2}$
Pieze	pascal	1.0×10^{3}
Pint (*dry*)	bushel	1.5625×10^{-2}
Pint (*liquid; US*)	cm³	4.732×10^{2}
Pint (*liquid, UK*)	ft³	$2.006\,79 \times 10^{-2}$
Pint (*liquid, US*)	ft³	$1.671\,02 \times 10^{-2}$
Pint (*liquid, UK*)	gallon(*UK*)	0.125
Pint (*liquid, UK*)	gallon(*US*)	0.150 118
Pint (*liquid, US*)	gallon (*UK*)	0.104 085
Pint (*US*)	gallon (*US*)	0.125
Pint	gill	4.0
Pint (*dry*)	in.³	33.6
Pint (*liquid, UK*)	in.³	34.6774
Pint (*liquid, US*)	in.³	28.8752
Pint (*liquid, UK*)	litre	0.568 245
Pint (*liquid, US*)	litre	0.473 166
Pint (*dry, US*)	metre³	5.506×10^{-4}
Pint (*liquid; US*)	metre³	4.732
Pint (*US*)	ounce	16.0
Pint (*liquid, UK*)	pint (*liquid, US*)	1.200 94
Pint (*liquid, US*)	pint (*liquid, UK*)	0.832 68
Pint (*dry*)	quart (*dry*)	0.5
Pint (*liquid*)	quart (*liquid*)	0.5
Pint (*liquid*)	yard³	6.189×10^{-4}
Planck's quantum	erg-sec	6.624×10^{-27}
Poise (*absolute viscosity*)	gram(mass)/cm-sec	1.0
Poise	kgf-sec/metre²	$1.019\,72 \times 10^{-2}$
Poise (*absolute viscosity*)	pascal-sec	0.1
Poise	poundf-sec/ft²	$2.008\,54 \times 10^{-3}$

To Convert	*into*	*multiply by*
Poise	poundal-sec/ft^2	6.7197×10^{-2}
Pole	rod	1.0
Pound (*avoirdupois*)*	cwt (*hundred weight; long*)	8.9286×10^{-3}
Pound (*avoirdupois*)	drachm (*apothecaries*)	$1.1\overline{6}$** $\times 10^2$
Pound (*avoirdupois*)	dram (*avoirdupois*)	2.56×10^2
Pound (*avoirdupois*)	dyne	$4.448\,23 \times 10^5$
Pound (*avoirdupois*)	grain	7.0×10^3
Pound (*avoirdupois*)	gram	$4.535\,923\,7 \times 10^2$
Pound (*avoirdupois*)	joule/metre (*newton*)	4.448
Pound (*avoirdupois*)	kilogram	0.453 592 37
Pound (*avoirdupois*)	litre (*water*)	0.453 597
Pound (*avoirdupois*)	maund	$1.215\,28 \times 10^{-2}$
Pound (*avoirdupois*)	metric technical weight	$4.625\,36 \times 10^{-2}$
Pound (*avoirdupois*)	ounce (*avoirdupois*)	16.0
Pound (*avoirdupois*)	ounce (*troy*)***	14.583 312
Pound (*avoirdupois*)	pound (*apothecaries*)	1.215 28
Pound (*avoirdupois*)	poundal	32.174
Pound (*avoirdupois*)	quarter (*avoirdupois*)	4.0×10^{-2}
Pound (*avoirdupois*)	seer	0.486 111
Pound (*avoirdupois*)	slug	3.1081×10^{-2}
Pound (*avoirdupois*)	tola	38.888 741
Pound (*avoirdupois*)	ton (*long*)	$4.464\,29 \times 10^{-4}$
Pound (*avoirdupois*)	ton (*short*)	5.0×10^{-4}

**** in USA, pound is defined in avoirdupois units***

***** symbol — over a digit indicate recurring decimal digit***

****** troy or apothecaries***

To Convert	*into*	*multiply by*
Pound (*troy*)	cwt	7.3469×10^{-3}
Pound (*troy*)	grain	5.76×10^{3}
Pound (*troy*)	gram	$3.732\,417\,7 \times 10^{2}$
Pound	hundred weight (*short*)	1.0×10^{-2}
Pound (*troy*)	ounce (*avoirdupois*)	13.1657
Pound (*troy*)	ounce (*troy*)	12.0
Pound (*troy*)	pennyweight (*troy*)	2.4×10^{2}
Pound (*troy*)	pound (*avoirdupois*)	0.822 857
Pound (*troy; apothecaries*)	seer	0.4
Pound (*troy*)	ton (*long*)	$3.673\,47 \times 10^{-4}$
Pound (*troy*)	ton (*metric*)	3.7324×10^{-4}
Pound (*troy*)	ton (*short*)	$4.114\,29 \times 10^{-4}$
Pound force	joule/cm	$4.448\,399 \times 10^{-2}$
Pound force	megadyne	0.444 822 05
Pound force	newton (*N*)	4.448 222
Pound of water	ft³	1.602×10^{-2}
Pound of water	gallon (*UK*)	9.98×10^{-2}
Pound of water	gallon (*US*)	0.1198
Pound of water	in.³	27.68
Pound of water (*evaporated from and at 212⁰F*)	kilowatt-hour	0.283 286 12
Pound of water (*raised from 62⁰ to 212⁰F*)	kilowatt-hour	$4.395\,604 \times 10^{-2}$
Pound of water	metre³	$4.535\,924 \times 10^{-4}$
Pound of water/minute	ft³/minute	$1.601\,794 \times 10^{-2}$
Pound of water/minute (*discharge*)	ft³/sec	2.67×10^{-4}

To Convert	*into*	*multiply by*
Pound/acre (*application rate, fertilizer, pesticide, crop yield etc.*)	kg/hectare	1.120 851
Pound-feet (*moment of force, torque, bending moment*)	cm-dyne	$1.355\ 818 \times 10^{7}$
Pound-feet	cm-gram	1.3825×10^{4}
Pound-feet	metre-kg	0.138 25
Pound-feet	newton-metre	1.355 818
Pound/feet (*beam loading, running weight*)	kg/metre	1.488 164
Pound/feet	newton/metre	14.5939
Pound mass/ft-sec	centipoise	$1.488\ 095\ 2 \times 10^{3}$
Pound mass/ft-sec	Pa-sec	1.488
Pound-feet^2 (*mass moment of inertia*)	kg-metre^2	$4.214\ 011 \times 10^{-2}$
Pound/feet^2 (*pressure*)	atmosphere	$4.725\ 41 \times 10^{-4}$
Pound/feet^2	bar	$4.788\ 033\ 7 \times 10^{-4}$
Pound/feet^2	cm of mercury	$3.590\ 664 \times 10^{-2}$
Pound/feet^2	ft of water	1.602×10^{-2}
Pound/feet^2	gram/cm^2	0.488 257 41
Pound/feet^2	in. of mercury	1.4139×10^{-2}
Pound/feet^2	in. (*of water, 4°C*)	0.192 222 67
Pound/feet^2	kg/cm^2	$4.882\ 431\ 1 \times 10^{-4}$
Pound/feet^2	kg/metre^2	4.882 428
Pound/feet^2	kilopascal (*kPa*)	$4.788\ 026 \times 10^{-2}$
Pound/feet^2	millibar	0.478 802 6
Pound/feet^2	mm of mercury	0.359 131
Pound/feet^2	pound/in.^{2}	6.944×10^{-3}
Pound-feet^2/sec (*angular momentum*)	kg-metre^2/sec	$4.214\ 011 \times 10^{-2}$
Pound/feet^3 (*density*)	gram/cm^3	$1.601\ 846\ 3 \times 10^{-2}$

To Convert	*into*	*multiply by*
Pound/feet^3	gram/litre	16.018 709 85
Pound/feet^3	kg/metre^3	16.018 463
Pound/feet^3	pound/gallon (*UK*)	0.160 544 57
Pound/feet^3	pound/in.^{3}	5.787×10^{-4}
Pound/feet^3	pound/mil-ft	5.456×10^{-9}
Poundf-ft/degree (*spring rate, torsional*)	newton-metre/ degree	1.355 818
Pound/feet-hour (*viscosity*)	metre-newton-sec/metre^2	0.414
Pound-feet/sec (*momentum, linear*)	kg-metre/sec	0.138 255
Pound/gallon (*density*)	gram/cm^3	99.776
Pound/gallon (*UK*)	gram/litre	99.779
Pound/gallon (*liquid*)	kg/litre	0.119 826 4
Pound/gallon (*UK*)	pound/in.^{3}	$3.604\ 66 \times 10^{-3}$
Pound/gallon (*UK*)	pound/ft^3	6.2288
Pound mass/feet-hour	centipoise	0.413 223 14
Pound/thousand gallon	gram/litre	0.119 832 23
Pound/million gallon	grain/gallon (*US*)	$6.999\ 86 \times 10^{-3}$
Pound/million gallon	parts per million	0.119 832 23
Pound/(*hp-hour*)	gram/kWh	$6.082\ 774 \times 10^2$
Pound/(hp-hour) (*specific fuel consumption*)	gram/MJ	$1.689\ 659 \times 10^2$
Pound/inch	gram/cm	1.786×10^2
Poundf-inch (*moment of force, torque, bending moment*)	N-m	0.112 984 8
Poundf/inch (*surface tension, spring rate*)	N/mm	0.175 126 8
Pound/inch^2 (*pressure*)	atmosphere	6.804×10^{-2}
Pound/inch^2	bar	6.8948×10^{-2}
Pound/inch^2	cm of mercury	5.170 630 82

To Convert	*into*	*multiply by*
Pound/inch^2	dyne/cm^2	6.896 552
Pound/inch^2	ft of water	2.307
Pound/inch^2	in. of mercury	2.036
Pound/inch^2	in. (*of water; 4^{0}C*)	27.677 830 06
Pound/inch^2	kg/cm^2	7.0307×10^{-2}
Pound/inch^2	kg/metre^2	7.0307×10^{2}
Pound/inch^2	kilopascal (*kPa*)	6.894 757
Pound/inch^2	metre of water	0.704
Pound/inch^2	mm of mercury	51.652 892 56
Pound/inch^2	ounce/in.^{2}	16.0
Pound/inch^2	pascal	$6.894\ 757 \times 10^{3}$
Pound/inch^2	pound/ft^2	1.44×10^{2}
Pound/inch^2	tonne (short)/ft^2	5.0×10^{-4}
Pound/inch^2	tonnef/in.^{2}	$4.464\ 29 \times 10^{-4}$
Pound/inch^3	gram/cm^3	27.6799
Pound/inch^3	gram/millilitre	27.6807
Pound/inch^3	kg/metre^3	$2.767\ 99 \times 10^{4}$
Pound/inch^3	pound/ft^3	1.728×10^{3}
Pound/inch^3	pound/gallon(*UK*)	$2.774\ 19 \times 10^{2}$
Pound/inch^3	pound/mil-ft	9.425×10^{-6}
Pound-in.^{2} -^{0}F	calorie/cm^3 -^{0}C (*volumetric specific heat*)	$2.958\ 579\ 88 \times 10^{-3}$
Pound/mil-ft	gram/cm^3	$2.936\ 857\ 6 \times 10^{6}$
Pound/mil-ft	kilogram/metre^3	$2.936\ 857\ 6 \times 10^{9}$
Pound/mil-ft	pound/ft^3	$1.832\ 844\ 6 \times 10^{8}$
Pound/sec	tonne (*metric*)/hour	1.632 919 66
Pound-sec/ft (*damping coefficient*)	newton-sec/metre	14.5939
Pound-sec/ft^2	kgf-sec/metre^2	4.882 43

To Convert	*into*	*multiply by*
Pound-sec°F	calorie/cm² -sec-°C/cm (*thermal conductivity*)	$1.915\ 709 \times 10^{-2}$
Pound of water/hour	tonne of water /24 houı	$1.200\ 005 \times 10^{-2}$
Pound/yard	kg/metre	0.496 054 7
Pound/yard³ (*density; soil, building material*)	kg/metre³	0.593 276 3
Poundal	dyne	$1.382\ 655 \times 10^{4}$
Poundal	gram	14.1
Poundal	joule/cm	1.383×10^{-3}
Poundal	joule/metre (*newton*)	0.138 255
Poundal	kilogram force	$1.409\ 81 \times 10^{-2}$
Poundal	megadyne	$1.382\ 552 \times 10^{-2}$
Poundal	pound	3.1081×10^{-2}
Poundal/ft²	bar	$1.488\ 161\ 7 \times 10^{-5}$
Poundal-sec/ft²	kgf-sec/metre²	0.151 751
Poundal-sec/ft²	poundf-sec/ft²	3.1081×10^{-2}
Poundal-sec/ft²	poise	14.8816
Poundf-sec/ft²	centipoise	$4.786\ 979\ 4 \times 10^{4}$
Poundf-sec/ft² (*absolute viscosity*)	Pa-sec	47.8803
Poundf-sec/ft²	poise	$4.788\ 03 \times 10^{2}$
Poundf-sec/ft²	poundal-sec/ft²	32.174

Q

To Convert	*into*	*multiply by*
Quadrant (*angle*)	degree	90.0
Quadrant (*angle*)	minute	5.4×10^3
Quadrant (*angle*)	radian	1.571
Quadrant (*angle*)	revolution	0.25
Quadrant (*angle*)	second	3.24×10^5
Quantum	einstein (*photon energy*)	$1.660\ 565\ 5\times10^{-24}$
Quart (*dry*)	barrel (*US; dry*)	$9.523\ 81 \times 10^{-3}$
Quart (*dry*)	bushel	1/32
Quart (*liquid*)	cm³	9.464×10^2
Quart (*fluid*)	cup	4.0
Quart (*liquid*)	ft³	3.342×10^{-2}
Quart (*liquid*)	gallon	0.25
Quart (*dry*)	in.³	67.2
Quart (*liquid*)	in.³	57.75
Quart (*dry*)	litre	1.1012
Quart (*liquid*)	litre	0.946 352 9
Quart (*liquid*)	metre³	9.464×10^{-4}
Quart (*fluid*)	ounce	32.0
Quart (*dry*)	peck	0.125
Quart (*dry*)	pint (*dry*)	2.0
Quart (*liquid*)	pint (*liquid*)	2.0
Quart (*liquid*)	yard³	1.238×10^{-3}
Quarter (*avoirdupois*)	cwt (*avoirdupois*)	0.25
Quarter (*length, cloth*)	nail	4.0
Quarter (*avoirdupois*)	pound (*avoirdupois*)	25.0
Quarter	yard	0.25
Quintal (*q*)	kilogram	1.0×10^2
Quintal	tonne	0.1

R

To Convert	*into*	*multiply by*
Rad (*absorbed dose*)	gray (*J/kg*)	1.0×10^{-2}
Radian	circumference	0.159 159 64
Radian	degree	57.2958
Radian	gon* (*g*)	63.7
Radian	grade	63.653 723 74
Radian	metric minute** (*c*)	6.37×10^{3}
Radian	metric second*** (*cc*)	6.37×10^{5}
Radian (*surveyor's means*)	mil (*angular*)	1.0186×10^{3}
Radian	minute	$3.437\ 75 \times 10^{3}$
Radian	quadrant	0.636 62
Radian	revolution	0.1591
Radian	sec	$2.062\ 65 \times 10^{5}$
Radian/second	degree/sec	57.296
Radian/second	revolution/minute	9.5493
Radian/second	revolution/sec	0.159 155
Radian/second/second	revolution/minute /minute	5.73×10^{2}
Radian/second/second	revolution/minute /sec	9.549
Radian/second/second	revolution/sec/sec	0.1592
Rankin (*temperature, R*)	kelvin (*K*)	5/9
Revolution	degree	3.6×10^{2}
Revolution	minute (*angular*)	$2.159\ 827\ 2 \times 10^{4}$
Revolution	quadrant	4.0
Revolution	radian	6.283
Revolution	sec	1.296×10^{6}

* *gon : a right angle divided into 100 parts each of which is called a gon (g)*

** *one gon consists of 100 metric minutes (1^c)*

*** *one gon consists of 10^4 metric seconds (1^{cc}); $1^g = 10^{2c} = 10^{4cc}$*

To Convert	*into*	*multiply by*
Revolution/minute	degree/sec	6.0
Revolution/minute	radian/sec	0.104 72
Revolution/minute	revolution/sec	$1.6\overline{6}^{*} \times 10^{-2}$
Revolution/minute/minute	radian/sec/sec	1.745×10^{-3}
Revolution/minute/minute	revolution/minute /sec	$1.6\overline{6}^{*} \times 10^{-2}$
Revolution/minute/minute	revolution/sec/sec	2.778×10^{-4}
Revolution/minute/sec	radian/sec/sec	0.104 723 01
Revolution/minute/sec	revolution/minute /minute	60.0
Revolution/minute/sec	revolution/sec/sec	$1.6\overline{6}^{*} \times 10^{-2}$
Revolution/second	degree/sec	3.6×10^{2}
Revolution/second	radian/sec	6.2832
Revolution/second	revolution/minute	60.0
Revolution/second/second	radian/sec/sec	6.2832
Revolution/second/second	revolution/minute /minute	3.6×10^{3}
Revolution/second/second	revolution/minute /sec	60.0
Right solid angle	solid angle of complete sphere	0.125
Right solid angle	square degree	5.16×10^{3}
Right solid angle	steradian (*sr*)	1.57
Rod (*surveyor's*)	chain (*gunther's*)	0.25
Rod	dekametre	0.502 916 92
Rod	fathom	2.75
Rod	ft	16.5
Rod	furlong	2.5×10^{-2}
Rod	in.	1.98×10^{2}
Rod	link	25.0

* *symbol — over a digit indicate recurring decimal digit*

To Convert	*into*	*multiply by*
Rod	metre	5.029 21
Rod	perch*	1.0
Rod	pole*	1.0
Rod	yard	5.5
Rood	acre	0.25
Rood	chain^2	2.5
Rood	hectare	0.101 171
Rood	mile^2	$3.906\ 25 \times 10^{-4}$
Rood	rod^2	40.0
Rood	yard^2	1.21×10^{3}
Rod^2	acre	6.25×10^{-3}
Rod^2	chain^2	6.25×10^{-2}
Rod^2	ft^2	2.7225×10^{2}
Rod^2	hectare	$2.529\ 29 \times 10^{-3}$
Rod^2	mile^2	9.7656×10^{-6}
Rod^2	rood	2.5×10^{-2}
Rod	yard	30.25

* *"perch" and "pole" were the earlier names of "rod"*

S

To Convert	*into*	*multiply by*
Savart (*frequency interval*)	cent	3.98
Savart (*natural*)	frequency interval (*Seventh, HCB tone*)	3.663×10^{-3}
Savart (*tempered*)	frequency interval (*Seventh, HCB tone*)	$3.623\ 19 \times 10^{-3}$
Savart (*natural*)	musical Interval (*major tone sixth,A*)	$4.506\ 53 \times 10^{-3}$
Savart (*tempered*)	musical Interval (*major tone sixth,A*)	4.4287×10^{-3}
Savart (*natural*)	musical Interval (*major tone, D*)	$1.956\ 947 \times 10^{-2}$
Savart (*tempered*)	musical Interval (*major tone, D*)	$1.992\ 032 \times 10^{-2}$
Savart (*natural*)	musical Interval (*major tone third, E*)	$1.031\ 992 \times 10^{-2}$
Savart (*tempered*)	musical Interval (*major tone third, E*)	$9.960\ 16 \times 10^{-3}$
Savart (*natural*)	musical Interval (*fourth, F- tone*)	8.0×10^{-3}
Savart (*tempered*)	musical Interval (*fourth, F- tone*)	$7.974\ 48 \times 10^{-3}$
Savart (*natural*)	musical Interval (*fifth, G- tone*)	$5.678\ 59 \times 10^{-3}$
Savart (*tempered*)	musical Interval (*fifth, G- tone*)	$5.694\ 76 \times 10^{-3}$
Savart (*frequency interval, natural*)	octave	3.32×10^{-3}
Savart(*frequency interval, natural*)	octave	3.32×10^{-3}

To Convert	*into*	*multiply by*
Savart (*natural*)	unison (*musical interval, C-tone*)	$6.644\ 52 \times 10^{-3}$
Savart (*tempered*)	unison (*musical interval, C-tone*)	$6.644\ 52 \times 10^{-3}$
Second (*angle*)	degree	$2.7\overline{7}^* \times 10^{-4}$
Second (*angle*)	minute	$1.6\overline{6}^* \times 10^{-2}$
Second (*angle*)	quadrant	$3.086\ 42 \times 10^{-6}$
Second (*angle*)	radian	4.848×10^{-6}
Second (*angle*)	revolution	7.716×10^{-7}
Second (*angular*)	mil (*angular*)	4.9383×10^{-3}
Second (*time*)	arc	$4.16\overline{6}^* \times 10^{-3}$
Second (*time*)	day	1.157×10^{-5}
Second (time)	hour	$2.7\overline{7}^* \times 10^{-4}$
Second (time)	minute	$1.6\overline{6}^* \times 10^{-2}$
Seventh (*H(B) tone, frequency interval*)	cent (*natural*)	1.088×10^{3}
Seventh (*H(B) tone, frequency interval*)	cent (*tempered*)	1.1×10^{3}
Seventh (*H(B) tone, frequency interval*)	savart (*natural*)	2.73×10^{2}
Seventh (*H(B) tone, frequency interval*)	savart (*tempered*)	2.76×10^{2}
Seventh (*H(B) tone, frequency interval*)	unison (*C tone, natural*)	15/8
Scruple	dram	$0.3\overline{3}^*$
Scruple	grain	20.0
Sear	kilogram	0.933 105 66
Sear	mound	2.5×10^{-2}
Sear	pound (*avoirdupois*)	2.057 143 33
Sear	tonne (*long*)	$9.18\ 366\ 41 \times 10^{-4}$

* *symbol — over a digit indicate recurring decimal digit*

To Convert	*into*	*multiply by*
Sear	tonne (*short*)	1.02857×10^{-3}
Siemen (*S*)	Mho	1.0
Siemen/metre	Mho/cm	1.0×10^{-2}
Siemen/metre	Mho/ft	0.304 799 99
Sign	degree	30.0
Slug	kg	14.593 903
Slug	metric technical weight	1.488 16
Slug	pound	32.174
Slug/ft^3	kg/m^3	$5.153\ 79 \times 10^{2}$
Slug/ft-sec	Pa-sec	47.88
Solid angle of complete sphere	right solid angle	8.0
Solid angle of complete sphere	square degree	4.13×10^{4}
Solid angle of complete	steradian	12.6
Span	in.	9.0
Sphere	hemisphere	2.0
Sphere	spherical right angle	8.0
Sphere	steradian (sr)*	12.57
Spherical candle power	lumen	12.565 971 35
Spherical right angle	hemisphere	0.25
Spherical right angle	sphere	0.125
Spherical right angle	steradian	1.571
Square degree	right solid angle	1.94×10^{-4}
Square degree	solid angle of complete sphere	2.42×10^{-5}
Square degree	steradian	3.05×10^{-4}
Statampere (*electric current*)	abampere	3.335×10^{-11}

* *4π steradians equal to 1 sphere*

To Convert	*into*	*multiply by*
Statampere	ampere	3.335×10^{-10}
Statcoulomb	abcoulomb	3.335×10^{-11}
Statcoulomb	ampere-hour	9.259×10^{-14}
Statcoulomb	coulomb	3.335×10^{-10}
Statcoulomb	faraday	3.457×10^{-15}
Statfarad (*capacitance*)	abfarad	1.112×10^{-21}
Statfarad	farad	1.112×10^{-12}
Stathenrie (*inductance*)	abhenrie	8.988×10^{20}
Stathenrie	henrie	8.988×10^{11}
Statohm (*electric resistance*)	abohm	8.988×10^{20}
Statohm	ohm	8.988×10^{11}
Statvolt (*electromotive force*)	abvolt	2.998×10^{10}
Statvolt	volt	2.998×10^{2}
Step (*military*)	ft	2.5
Stere* (*volume*)	bushel	28.375
Steradian (*solid angle, sr*)	hemisphere	0.1592
Steradian	solid angle of complete sphere	$7.936\,508 \times 10^{-2}$
Steradian	sphere	7.958×10^{-2}
Steradian	spherical right angle	0.6366
Steradian	square degree	3.28×10^{3}
Stilb (*luminance*)	apostilb	3.14×10^{4}
Stilb	lambert	3.14
Stilb (*luminance, photometric brightness, cd/cm*^{2})	nit (cd/m^2)	1.0×10^{4}
Stoke	ft^2/hour	3.875 007 75
Stoke (*kinematic viscosity*)	ft^2/sec	$1.076\,39 \times 10^{-3}$
Stoke	$metre^2$/sec	1.0×10^{-4}

* *"stere" is the unit of volume equal to 1 kilolitre*

T

To Convert	*into*	*multiply by*
Tablespoon	$metre^3$	1.479×10^{-5}
Teaspoon	$metre^3$	4.929×10^{-6}
Temperature (*°C*)	°F	°F=[1.8 × °C+32]
Temperature (*°C*)	kelvin (*K*)	K=°C+273.15
Temperature (*°C*)	rankin (*R*)	R=(°C+273.15) × 9/5
Temperature (*°C+273.15*)	absolute temperature, kelvin (*K*)	1.0
Temperature (*°F*)	°C	°C=[(°F-32)×5/9]
Temperature (*°F*)	kelvin (*K*)	K=(°F+459.67) ×5/9
Temperature (*°F*)	rankin (*R*)	R=°F+459.67
Temperature (*°F+459.67*)	absolute temperature, rankin (*R*)	1.0
Temperature (*kelvin, K*)	°C	°C=(K-273.15)
Temperature (*kelvin, K*)	°F	°F=(1.8 × K-459.67)
Temperature (*kelvin, K*)	R	9/5
Temperature (*rankin, R*)	°C	°C=(R × 5/9-273.15)
Temperature (*rankin, R*)	°F	°F=(R-459.67)
Temperature (*rankin, R*)	kelvin (*K*)	5/9
Tesla (*T, magnetic induction*)	cgse	$3.3\overline{3}$* $\times 10^{-7}$
Tesla	gauss	1.0×10^{4}
Tesla	kilogauss	10.0
Therm	megajoules (*MJ*)	1.055×10^{2}
Tola	carat (*metric*)	58.319
Tola	drachm (*apothecarie's*)	3.0

* *symbol — over a digit indicate recurring decimal digit*

To Convert	*into*	*multiply by*
Tola	dram (*apothecarie's*)	6.5829
Tola	grain	1.8×10^2
Tola	gram	11.6638
Tola	ounce (*apothecarie's*)	0.375
Tola	ounce (*avoirdupois*)	0.411 429
Tola	pound	$2.571\ 438 \times 10^{-2}$
Ton (*refrigeration*)	kilowatt	3.517
Tonne	quintal	10.0
Tonne force	kilonewton (*kN*)	9.964
Tonne/in.^{2}	pound/in.^{2}	$2.239\ 997\ 8 \times 10^3$
Tonne (*long*)	hundred weight (*long; cwt*)	20.0
Tonne (*long*)	hundred weight (*short*)	22.399 978 5
Tonne (*long, 2240 lbm*)	kilogram	$1.016\ 047 \times 10^3$
Tonne (*long*)	maund*	$27.22\bar{2}$**
Tonne (*long*)	ounce	$3.584\ 229\ 4 \times 10^4$
Tonne (*long*)	pound	2.24×10^3
Tonne (*long*)	pound (*troy*)	$2.722\ 221\ 8 \times 10^3$
Tonne (*long*)	seer*	$1.088\ 89 \times 10^3$
Tonne (*long*)	tonne (*short*)	1.12
Tonne (*metric*)	hundred weight (*short*)	22.046 244 2
Tonne (*metric*)	kilogram	1.0×10^3
Tonne (*metric*)	ounce	$3.527\ 336\ 9 \times 10^4$
Tonne (*metric*)	pound	2.205×10^3
Tonne (*metric*)	tonne (*short*)	1.102 311 36

* *used in India*

** *symbol — over a digit indicate recurring decimal digit*

To Convert	*into*	*multiply by*
Tonne (*register*)*	ft^3	1.0×10^2
Tonne (*shipping, UK, cargo*)	bushel	32.703
Tonne (*shipping, UK*)	ft^3	42.0
Tonne (*shipping, US*)	ft^3	40.0
Tonne (*shipping, US, cargo*)	bushel	32.143
Tonne (*short*)**	cwt	17.8571
Tonne (*short, 2000 lbm*)	kilogram	$9.071\,848 \times 10^2$
Tonne (*short*)	maund	24.3056
Tonne (*short*)	ounce	3.2×10^4
Tonne (*short*)	ounce (*troy*)	$2.916\,666 \times 10^4$
Tonne (*short*)	pound	2.0×10^3
Tonne (*short*)	pound (*troy*)	$2.430\,56 \times 10^3$
Tonne (*short*)	seer	9.7222×10^2
Tonne (*short*)	tonne (*long*)	0.892 87
Tonne (*short*)	tonne (*metric*)	0.907 184 7
Tonne/acre	kilogram/hectare	2.511×10^3
Tonne/ft^2	atmosphere	0.945 179 58
Tonne/in.^{2}	atmosphere	$1.360\,914\,5 \times 10^2$
Tonne/in.^{2}	bar	$1.544\,43 \times 10^2$
Tonne/in.^{2}	kilogramf/cm^2	$1.574\,88 \times 10^2$
Tonne/hectare	tonne (short)/acre	0.446 089 62
Tonne (*UK*)/hour	kg/sec	0.282 235 5
Tonne/hour (megagram/ hour)	tonne (*short*)/hour	1.102 311 36
Tonne (*metric*)/hour	kilogram/sec	$0.27\overline{7}$***
Tonne (*metric*)/hour	pound/sec	0.6124

* *for register tonnage or measurement of the entire internal capacity of a vessel, it is arbitrarily assumed, to facilitate computation*

** *coal production is measured in terms of short tonne*

*** *symbol — over a digit indicate recurring decimal digit*

To Convert	*into*	*multiply by*
Tonne (*short*)/acre (*crop yield, soil erosion*)	tonne/hectare	2.241 702
Tonne(*short*)/ft^2	kilogram/metre^2	9.765×10^3
Tonne(short)/ft^2	pound/in.^{2}	2.0×10^3
Tonne (*short*)/hour *machine work capacity, harvesting, material handling*)	tonne/hour (*megagram/hour*)	0.907 184 7
Tonne of water/24 hour	ft^3/hour	1.3349
Tonne of water/24 hour	gallon/minute	0.166 43
Tonne of water/24 hour	pound of water /hour	83.333

U

To Convert	*into*	*multiply by*
Unison (*musical interval, C tone, natural*)	cent (*natural scale*)	6.0×10^2
Unison	cent (*tempered*)	6.0×10^2
Unison	musical interval (*major tone,sixth, A*)	3/5
Unison	musical interval (*major tone, D*)	8/9
Unison	musical interval (*major tone,third, E*)	4/5
Unison	musical interval (*fourth, F-tone*)	3/4
Unison	musical interval (*fifth, G-tone*)	2/3
Unison	octave	0.5
Unison	savart (*natural scale*)	1.505×10^2
Unison	savart (*tempered*)	1.505×10^2
Unison	seventh (HCB) tone (*frequency interval*)	8/15

V

To Convert	*into*	*multiply by*
Varas	metre	0.848 176 42
Volt (*electromotive force*)	abvolt	1.0×10^{8}
Volt (*potential difference*)	cgse	$3.3\overline{3}^{*} \times 10^{-3}$
Volt	cgsm	1.0×10^{8}
Volt (*International*)	joule	9.654×10^{4}
Volt (*International*)	joule (*absolute*)	1.593×10^{-19}
Volt (*absolute*)	statvolt	3.336×10^{-3}
Volt (*International*)	volt (*absolute*)	1.0003
Volt (*absolute*)	volt (*International*)	0.999 700 09
Volt/in.	volt/cm	0.3937
Volt/metre (*field intensity*)	cgse	$3.3\overline{3}^{*} \times 10^{-5}$
Volt/metre	cgsm	1.0×10^{6}

**** symbol — over a digit indicate recurring decimal digit***

W

To Convert	*into*	*multiply by*
Watt	Btu/hour	3.412 14
Watt	calorie/sec	0.238 846
Watt	erg/sec	1.0×10^{7}
Watt	foot-pound/minute	44.27
Watt	foot-pound/sec	0.7378
Watt	hectowatt	1.0×10^{-2}
Watt	horsepower	1.341×10^{-3}
Watt	horsepower (*metric*)	1.36×10^{-3}
Watt	kilocalorie/hour	0.859 85
Watt	kilowatt	1.0×10^{-3}
Watt	lumen	$6.684\ 492 \times 10^{2}$
Watt	metre-kilogram /sec	0.102
Watt (*absolute*)	Btu(*mean*)/minute	5.6884×10^{-2}
Watt (*absolute*)	joule/sec	1.0
Watt (*International*)	watt (*absolute*)	1.0002
Watt/cm^2	Btu/ft^2-hour	$3.169\ 98 \times 10^{3}$
Watt/cm^2	calorie/cm^2-sec	0.238 846
Watt/cm^2	kilocalorie/m^2-hour	$8.598\ 452\ 28 \times 10^{3}$
Watt/cm-°C	Btu/ft-hour-°F (*thermal conductivity*)	57.779
Watt/cm-°C	Btu-in./ft^2-hour-°F	6.9335×10^{2}
Watt/cm-°C	calorie/cm-sec-°C	0.238 846
Watt/cm-°C	kilocalorie/metre-hour-°C	85.985
Watt/cm^2-°C (*surface heat transfer coefficient*)	Btu/ft^2-hour-°F	1.7611×10^{3}

To Convert	*into*	*multiply by*
Watt/cm²-°C	calorie/cm²-sec-°C	0.238 846
Watt/cm2-°C	kilocalorie/metre²-hour-°C	8.5985×10^3
Watt-hour	Btu	3.413
Watt-hour	erg	3.6×10^{10}
Watt-hour	ft-pound	2.656×10^3
Watt-hour	gram-calorie	8.5985×10^2
Watt-hour	horsepower-hour	1.341×10^{-3}
Watt-hour	joule	3.6×10^3
Watt-hour	kilogram-calorie	0.8605
Watt-hour	kilogram-metre	3.672×10^2
Watt/in.²	Btu/ft/min (*solar irradiance, insolation*)	8.190 008 19
Watt/m-K	Btu-ft/(h-ft²-°F)	0.577 789 21
Watt/m-K	Btu-in./ft²-h-°F	6.939 625 26
Watt/m²	Btu/ft²-h (*solar radiation*)	0.316 998 31
Watt/m²	Btu/ft²/min	$5.286\,840\,2 \times 10^{-3}$
Watt/m²	Btu/(h-ft²) (*density of heat flow rate, irradiance*)	0.316 998 31
Watt/m²	erg/sec-cm² (*sound intensity*)	1.0×10^3
Watt/m²	erg	1.0×10^3
Watt/m²	langley/min	$1.434\,03 \times 10^{-3}$
Watt/(m²-K)	Btu/(h-ft²-°F)	0.176 110 19
Watt-second	joule	1.0
Weber (*magnetic flux, Wb*)	cgse	$3.3\overline{3}^* \times 10^{-3}$
Weber	kiloline	1.0×10^5
Weber	line	1.0×10^8

* *symbol — over a digit indicate recurring decimal digit*

To Convert	*into*	*multiply by*
Weber	maxwell	1.0×10^8
Weber/cm^2	gauss	1.0×10^8
Weber/cm^2	line/in.^{2}	$6.451\,612\,9 \times 10^8$
Weber/cm^2	Weber/in.^{2}	6.451 612 9
Weber/cm^2	Weber/m^2	1.0×10^4
Weber/inch^2	gauss	1.55×10^7
Weber/inch^2	line/in.^{2}	1.0×10^8
Weber/inch^2	weber/cm^2	0.155
Weber/inch^2	weber/metre^2	1.55×10^3
Weber/metre^2	gauss	1.0×10^4
Weber/metre^2	line/in.^{2}	6.452×10^4
Weber/metre^2	weber/cm^2	1.0×10^{-4}
Weber/metre^2	weber/in.^{2}	6.452×10^{-4}
Week	hour	1.68×10^2

Y

To Convert	*into*	*multiply by*
Yard	cable length	$4.16\overline{6}$* × 10^{-3}
Yard	chain	$4.5\overline{45}$* × 10^{-2}
Yard	cm	91.44
Yard	fathom	0.5
Yard	ft	3.0
Yard	furlong	4.545 45 × 10^{-3}
Yard	in.	36.0
Yard	kilometre	9.144 × 10^{-4}
Yard	metre	0.9144
Yard	mil	3.6 × 10^{4}
Yard	mile (*nautical, International*)	4.937 37 × 10^{-4}
Yard	mile (*statute*)	5.6818 × 10^{-4}
Yard	mile (*nautical, UK*)	4.934 21 × 10^{-4}
Yard	millimetre	9.144 × 10^{2}
Yard	quarter	4.0
Yard	rod	$0.\overline{18}$*
Yard/hour	knot	4.933 399 1 × 10^{-4}
$Yard^2$	acre	2.066 12 × 10^{-4}
$Yard^2$	are	8.3612 × 10^{-3}
$Yard^2$	bigha	3.305 785 124 × 10^{-4}
$Yard^2$	$chain^2$	2.066 12 × 10^{-3}
$Yard^2$	cm^2	8.361 × 10^{3}
$Yard^2$	ft^2	9.0
$Yard^2$	hectare	8.3613 × 10^{-5}
$Yard^2$	$in.^2$	1.296 × 10^{3}
$Yard^2$	$kilometre^2$	8.361 204 × 10^{-7}

** symbol — over a digit indicate recurring decimal digit*

To Convert	*into*	*multiply by*
Yard²	metre²	0.836 127 36
Yard²	mile²	$3.228\ 31 \times 10^{-7}$
Yard²	mm²	8.361×10^{5}
Yard²	rod²	$3.305\ 79 \times 10^{-2}$
Yard²	rood	8.2645×10^{-4}
Yard²/gallon (*UK, surface coating*)	metre²/litre	0.1839
Yard³	acre-ft	6.1983×10^{-4}
Yard³	barrel (*US, liquid*)	$4.808\overline{8}$*
Yard³	bushel (*UK*)	21.0223
Yard³	bushel (*US*)	21.6961
Yard³	cm³	7.646×10^{5}
Yard³	ft³	27.0
Yard³	gallon (*US*)	2.02×10^{2}
Yard³	in.³	4.6656×10^{4}
Yard³	kilolitre	0.764 53
Yard³	litre	7.6453×10^{2}
Yard³	metre³	0.764 554 858
Yard³	pint (*US*)	1.6159×10^{3}
Yard³	quart (*US, liquid*)	8.0795×10^{2}
Yard³/minute	ft³/sec	0.45
Yard³/minute	gallon(*US*)/sec	3.367
Yard³/minute	litre/sec	12.74

* *symbol — over a digit indicate recurring decimal digit*

Bibliography

1. Adams, L.F. *Engineering Measurements and Instrumentation.* The English Universities Press Ltd. St Pauls House, Warwick Lane, London EC4P 4AH. 1975.

2. Agricultural Engineers Yearbook - 1982. ASAE Engineering Practice: ASAE EP285.6. Use of SI (Metric) Units pp. 100-109.

3. *Algebra with Arithmetic and Mensuration from the Sanskrit of Brahmagupta and Bhascara,* translated by Henry Thomas Colebrooke, 1817, includes the 'Lilavati').

4. Clark's Tables, Science Data Book. Oliver & Boyd. Croythorn House, 23 Ravelston Terrace. Edinburgh EH4 3TJ, UK. PP : 37-53.

5. Eshbach, Ovid W. *Handbook of Engineering Fundamentals,* Vol. I of Wiley Engineering Handbook Series. John Wiley & Sons, Inc. New York 1936. pp. 1-130 to 1-147 & 3-02 to 3-36.

6. Fairbanks - Morse Enginjeering Tables (UK).

7. *Historical Metrology*, A.E. Berriman, 1953, J.M. Dent & Sons Ltd., New York.

8. Home Page of National Institute of Standards and Technology (NIST), [URL http:/www.nist.gov].

9. *ISI Handbook of Quantities, Conversion Factors, Formulae and Tables*. 1960. Indian Standards Institution, Manak Bhavan, 9, Mathura Road, New Delhi-1.

10. Verman and Jainath Kaul, *Metric Change in India*, Lal C. 1970, Indian Standards Institution, New Delhi.

11. Murphy, Glenn. *Similitude in Engineering*. The Ronald Press Company. New York. 1950 pp. 4-7.

12. Sena, L.A. *Units of Physical Quantities and their Dimensions*. MIR Publishers. Moscow 1972 (Translated from the Russian by G. Leib).

13. Weights and Measures. *The New Encyclopedia Britannica*. 15th Edition. Vol. 19, 1978. pp. 728-735.